KV-512-458

£6·95

Pocket Examiner
in
# Biochemistry

# Pocket Examiner
## in
# Biochemistry

## Desmond G O'Sullivan

*Emeritus Reader in Chemistry
in Relation to Medicine in the
University of London,
formerly of the
Courtauld Institute of Biochemistry,
The Middlesex Hospital Medical School,
London*

# Churchill Livingstone

Edinburgh London Melbourne and New York 1986

CHURCHILL LIVINGSTONE
Medical Division of Longman Group Limited

Distributed in the United States of America by
Churchill Livingstone Inc., 1560 Broadway, New York,
N.Y. 10036, and by associated companies, branches
and representatives throughout the world.

© O'Sullivan, D. G.

All rights reserved. No part of this publication may be
reproduced, stored in a retrieval system, or transmitted in
any form or by any means, electronic, mechanical,
photocopying, recording or otherwise, without the prior
permission of the publishers (Churchill Livingstone, Robert
Stevenson House, 1–3 Baxter's Place, Leith Walk,
Edinburgh EH1 3AF).

First published 1983 (Pitman Publishing Ltd)
  Reprinted 1986 (Churchill Livingstone)

ISBN 0 443 03656 X

**British Library Cataloging in Publication Data**

O'Sullivan, D. G.
  Pocket examiner in biochemistry.
  1. Biological chemistry
  I. Title
  574.1'92'02461     QP514.2

**Library of Congress Cataloguing in Publication Data**

O'Sullivan, Desmond Gerard.
  Pocket examiner in biochemistry.
  Includes bibliographical references.
  1. Biological chemistry—Examinations,
questions, etc. I. Title.
QP518.5.08     574.19'2'076     82-5417

Printed at The Bath Press, Avon

# Contents

# Preface

The *Pocket Examiner* aims to help students in their revision studies and to provide practice in relation to oral examinations. Practice is very desirable as the biochemistry 'oral' frequently decides the fate of a student in this subject. Naturally the book is not intended to supplant textbooks and other sources. A student should refer at first to his own textbook for amplification of any answer in the *Pocket Examiner*. However, as no textbook is comprehensive, a textbook reference is given after most answers. The list of textbooks contains only editions with publication dates of 1978 or later and books that have undergone revision have been preferred to first edition books. Restriction of references to textbooks does not imply that students should restrict general course reading to textbooks only.

The book has been written with the needs of medical and dental students in mind but, in order to limit its size, the omission or severe restriction of some applied topics (e.g. digestion, immunochemistry, endocrinology and neurochemistry) has been necessary. These are, however, borderline subjects and are included in companion *Pocket Examiners.*

Chemical nomenclature in the *Pocket Examiner* is that normally employed in biochemical textbooks, thus 'acetic acid' and 'oxalic acid' are used rather than the systematic names 'ethanoic acid' and 'ethanedioic acid' respectively. Abbreviations, such as ATP and CoA, are standard biochemical abbreviations. In the *Pocket Examiner*, CoA is bracketed if the symbol excludes the $-SH$ group, e.g., in $CH_3CO.S(CoA)$. Very occasionally, clarity or brevity has been assisted by the inclusion of numerical data that a student would not be expected to remember.

A few questions in the *Pocket Examiner* require the performance of simple calculations. The examinee is unlikely to be confronted with such questions in oral examinations.

DGO'S

# 1 Key to References and Further Reading

A    *Biochemistry*, Bhagavan, N. V., 2nd edn. (J. B. Lippincott, Philadelphia, 1978)

B    *Biochemistry, a Functional Approach*, McGilvery, R. W., 2nd edn. (W. B. Saunders, Philadelphia, 1979)

C    *Harper's Review of Biochemistry*, Martin, D. W., Mayes, P. A. and Rodwell, V. W., 18th edn. (Lange Medical Publications, Los Altos, California, 1981)

D    *Biochemistry, a Case-Oriented Approach*, Montgomery, R., Dryer, R. L., Conway, T. W. and Spector, A. A., 3rd edn. (C. V. Mosby, St. Louis, 1980)

E    *Principles and Problems in Physical Chemistry for Biochemists*, Price, N. C. and Dwek, R. A., 2nd edn. (Clarendon Press, Oxford, 1979)

F    *Biochemistry*, Stryer, L., 2nd edn. (W. H. Freeman, San Francisco, 1981)

G    *Principles of Biochemistry*, White, A., Handler, P., Smith, E. L., Hill, R. L. and Lehman, I. R., 6th edn. (McGraw-Hill, New York, 1978)

H    *Basic and Applied Dental Biochemistry*, Williams, R. A. D. and Elliott, J. C. (Churchill Livingstone, Edinburgh, 1979)

I    *Clinical Chemistry in Diagnosis and Treatment*, Zilva, J. F. and Pannal, P. R., 3rd edn. (Lloyd-Luke (Medical Books), London, 1979)

# 2 Questions

## ACIDS AND BASES

### A Proton transfer and pH

1 What are the definitions of the terms 'acid' and 'base' that are employed in modern biochemistry and medicine?

2 When an acid is dissolved in water, an equilibrium is very rapidly attained. Bearing in mind that the acid may itself be an ion, write a general chemical equation that will represent this equilibrium.

3 The hydronium ion molarity $[H_3O^+]$ of an aqueous solution is often called its hydrogen ion concentration and is written as $[H^+]$. Using this convention, what mathematical equation relates the equilibrium concentrations of conjugate acid, conjugate base and $[H^+]$ in water at constant temperature?

4 Is the water molecule $(H_2O)$ itself regarded as an acid or as a base?

5 Give the equation that relates $[H^+]$ and $[OH^-]$ in water at constant temperature.

6 Classify the following as acids or bases: $H_2CO_3$, $CO_2$, $CO_3^{2-}$, $CN^-$, $NH_3$, $NH_4^+$, $OH^-$, $CH_3.CHOH.COO^-$, $CH_3.COOH$, $H_3O^+$, $H_3\overset{+}{N}.CH_2.COOH$.

7 Classify the following as acids or bases: $HCO_3^-$, $H_2PO_4^-$, $H_3\overset{+}{N}.CH_2.COO^-$.

8 The proportion of an acid that reacts with water in aqueous solution to give the conjugate base is called the degree of ionization $(\alpha)$.
How does $\alpha$ vary with dilution (l/mol) for a weak acid and for a strong acid?

9 Define the entities: pH, $pK_a$ and $pK_w$.

3

10 Calculate the pH of $10^{-3}$ mol/l solutions of (a) HCl, (b) $H_2CO_3$, (c) NaOH and (d) $NaHCO_3$. Take the (first) $pK_a$ of $H_2CO_3$ to have an effective value of 6.

# B Buffers

11 Describe what is meant by a buffer.

12 What are the components of a buffer solution?

13 What would be the important dissolved components of:
(a) an acetate buffer (of pH 5),
(b) a bicarbonate buffer (of pH 6),
(c) a phosphate buffer (of pH 7)?

14 State which of the following pairs, with molar proportions of 1/2 for each pair, will possess buffer action and give the conjugate acid and base pairs in each case:
(a) HCl/NaCl
(b) $HCl/HOCH_2CH_2NH_2$
(c) $HCl/H_3\overset{+}{N}.CH_2.COO^-$
(d) $H_3\overset{+}{N}.CH_2.COO^-/NaOH$

15 A buffer solution always contains a weak acid and its conjugate base. What equation is usually used to relate the concentrations of these components to the pH of the solution?

16 If the ratio [base]/[acid] has the values 0.1, 1.0, and 10, what are the respective values of the pH of this solution in terms of the $pK_a$ of the acid?

17 At which pH will a buffer solution be equally effective at resisting the pH change consequent upon the addition of a *small* quantity of either $H_3O^+$ or $OH^-$ ions?

18 How does the effectiveness of a buffer change with its pH?

19 Assuming that the ratio [base]/[acid] lies within the buffering range, what determines the capacity of the buffer to take up
(a) added $H_3O^+$ ions,
(b) added $OH^-$ ions?

20 Why is it necessary for blood to be buffered?

21 Describe briefly how buffering occurs in the blood.

4

## C  Acidosis and alkalosis

22  Give a form of the Henderson–Hasselbalch equation that is appropriate to the carbon dioxide–bicarbonate system in plasma.

23  Define what is meant by the terms acidosis and alkalosis.

24  Give examples of clinical conditions producing respiratory acidosis.

25  Give examples of clinical conditions producing respiratory alkalosis.

26  What happens to plasma $P_{CO_2}$, $[H^+]$ and $[HCO_3^-]$ in respiratory acidosis and in respiratory alkalosis?

27  What longer term processes will exert compensating effects on states of respiratory acidosis and respiratory alkalosis?

28  Give examples of underlying clinical circumstances that may be responsible for the development of non-respiratory (metabolic) acidosis.

29  Indicate how a state of non-respiratory (metabolic) alkalosis may arise.

30  What happens to $P_{CO_2}$ and to $H^+$ and $HCO_3^-$ concentrations
(a) in diabetic ketoacidosis;
(b) with repeated vomiting in pyloric stenosis?

31  What is the order in time in which the 3 different processes can operate to help offset the effect of an abnormal change in plasma $[H^+]$? Which of these processes act to remove $H^+$ ions from the body?

# AMINO ACIDS AND PEPTIDES

## A  Fundamental properties of amino acids

32  Which of the following conventional formulae represent amino acids (of any type):
(a) $H_2N.CH_2.COOH$

(b) $H_2N.CH_2.CH_2.COOH$
(c) $H_2N.CH_2.CH_2.SO_2OH$
(d) $H_2N.CO.CH_2.COOH$
(e) $H_2N.CH_2.CH_2.OH$
(f) $H_3C.NH.CH_2.COOH$

(g) $H_2N$—⬡—$COOH$

(h) $H_2N$—⬡—$SO_2.NH_2$

33 What general characteristics are possessed by the chemical structures of those 20 amino acids that commonly participate in protein structure?

34 Comment briefly on the molecular features that influence the solubilities in water of these 20 amino acids.

35 What type of ionic structure is possessed by the $\alpha$-amino acids when in the crystalline form and at the isoelectric point in solution?

36 Excluding glycine, all the amino acids that are protein constituents are L-amino acids. Explain this statement.

37 Which is the stronger acid, acetic acid or fully protonated glycine?

38 Alanine is considered to have 2 ionization constants. What are the respective ionization stages? If $pK_1 = 2.35$ and $pK_2 = 9.69$, what is the isoelectric point $pI$ of alanine?

39 Write down the ionization stages of fully protonated aspartic acid. How many $pK_a$ and $pI$ values exist?

40 Consider fully protonated lysine. As the pH is increased, what is the order in which the 3 acidic groups ionize? Give the dominant structure at the isoelectric point. [Lysine = 2,6-diaminohexanoic acid]

41 Can peptides be prepared by mixing together solutions of amino acids?

42 Name a test that is used to demonstrate the

6

presence of an amino acid (particularly on a paper chromatogram). How specific is the test for amino acids?

**B  Properties of amino acid side chains** [*The amino acids included are the* 20 *commonly present in proteins*]

43  Three letter abbreviations are used as symbols for certain amino acids. Apart from the amino acid molecules themselves, what in general do the 3 letter symbols represent? What are the recognized 3 letter symbols for (a) isoleucine, (b) asparagine, (c) glutamine and (d) tryptophan?

44  Why has it been found desirable for some purposes to have a second abbreviation scheme for these α-amino acids? Give four examples of these alternative abbreviations.

45  Give, in each case, 2 examples of α-amino acids with
(a)  acidic side chains,
(b)  side chains containing an amide group,
(c)  basic side chains,
(d)  side chains containing alcoholic OH groups,
(e)  side chains containing a sulphur atom.

46  What geometrical aspects of the molecules of (a) glycine and (b) proline give these 2 amino acids distinctive molecular properties?

47  Which 2 of the 20 α-amino acids possess a single ring containing 6 carbon atoms? How will these 2 differ in lipophilic properties?

48  Is histidine an acidic, basic or neutral amino acid?

49  What major distinctive property is possessed by the amino acid cysteine?

50  Select any 5 amino acids with markedly hydrophilic side chains and any 5 with markedly lipophilic side chains.

51  List 6 of the amino acids whose *side chains* can exhibit no hydrogen-bonding properties.

7

52 Which of the 20 amino acids
  (a) have more than one asymmetric carbon atom,
  (b) strongly absorb light at a wavelength above 270 nm in the near ultraviolet region of the spectrum.

# C The peptide bond and polypeptides

53 Define what is meant by a peptide bond.

54 Does a dipeptide molecule possess free rotation about the peptide bond as is usually the case with single bonds?

55 Does glutathione possess a normal tripeptide structure?

56 Why are the tetrapeptides Ala–Gly–Ser–Ala and Ala–Ser–Gly–Ala not identical?

57 Some polypeptides synthesized by fungi contain D-amino acids and also other amino acids that are not present in protein synthesized by mammalian cells. Gramicidin S is a cyclic polypeptide of fungal origin. Explain why the formula

L-Val–L-Orn–L-Leu–D-Phe–L-Pro
|                                               |
L-Pro–D-Phe–L-Leu–L-Orn–L-Val

does not uniquely define the structure of this antibiotic.

58 Describe how entirely covalent cross-links may be formed between 2 polypeptide chains or between 2 places in the same chain.

59 How may the disulphide bond in a cystine residue be broken?

60 What methods are available for the hydrolytic breakdown of polypeptides?

61 Animal thyrotropic regulatory hormone (TRH) has the structure:

Comment on this structure and decide which derivatives of which amino acids are present as units (or residues) in this structure.

# INVESTIGATION TECHNIQUES

## A   Separation of subcellular fractions

62   List some methods that have been used to disrupt animal and bacterial cells.

63   Is it enough to quote rotor speed and time of operation when describing the centrifugal field used in the separation of a subcellular fraction?

64   Describe briefly the process of differential centrifugation.

65   What is the order of sedimentation of the major cell components?

66   What is density gradient centrifugation?

67   Name solutes that are commonly used in preparing density gradients.

68   What is (rate)-zonal centrifugation?

69   What is isopycnic centrifugation?

70   What is equilibrium isodensity centrifugation?

71   Describe what is meant by the term 'marker enzyme' in relation to a particular subcellular structure.

## B   Separation of molecular components

72   State how the fundamental requirements for a suitable molecular separation method depend on whether the primary purpose of the separation is to isolate the components preparatively or to determine their percentage analytically.

73   List some separation methods that are particularly appropriate for preparative purposes and others that are particularly appropriate for analytical purposes.

74 Distinguish between adsorption and partition chromatography.

75 What is meant by the $R_F$ value for a component?

76 Outline briefly the principle involved in gas-liquid chromatography (g.l.c.).

77 Give briefly the principle involved in gel filtration or permeation chromatography.

78 Describe briefly the principle involved in affinity chromatography applied to the purification of an enzyme.

79 Outline the principle of ion exchange chromatography.

80 Distinguish between dialysis and ultrafiltration.

81 Outline the principle of electrophoretic separation.

82 Most electrophoresis is conducted with migration occurring in an inert supporting medium. Give examples of such media.

## C Colorimetry and spectrometry

83 Consider a parallel-flat-faced glass container (cell) containing an aqueous solution with a coloured solute. If $I_0$ is the intensity of incident light normal to the flat surface and if $I$ is the intensity of transmitted light normal to the parallel-flat surface, could the ratio $I_0/I$ be taken as a measure of the concentration of the coloured solute?

84 Define the Beer–Lambert Law.

85 Why are optical filters used in colorimeters?

86 What are 'blank' solutions and why are they used in colorimetry?

87 Give the relations between wavelength $\lambda$, frequency $\nu$ and wavenumber $\bar{\nu}$ for electromagnetic radiation.

88  What is the relation between the magnitude of an energy transition $\Delta E$ in a molecule or atom and the frequency $\nu$ of the electromagnetic radiation which is required to produce this transition?

89  What is the character of the energy transition in a molecule that gives rise to an absorption band in the visible or ultraviolet region of the spectrum?

90  What type of energy change in a molecule will be responsible for absorption in the near infrared region of the spectrum?

91  Give the origin of the light emission produced in a flame photometer.

92  Mention some other types of spectra of interest to biochemists.

## D  Radioisotopes

93  What do isotopes $^2H$, $^{12}C$, $^{13}C$, $^{14}N$ and $^{15}N$ have in common?

94  How does the radioactivity of $^{226}Ra$ differ from that of $^{14}C$ or $^{32}P$? Give some characteristic differences between the radioactivity of $^{14}C$ and $^{32}P$.

95  Give 4 radioactive isotopes that have been particularly useful as tracers in biochemical research and 4 that are useful in clinical diagnostic investigations.

96  How do $\gamma$-rays differ from X-rays?

97  Define the term microcurie.

98  What is meant by the half-life of a radioactive isotope?

99  Give 2 types of counters commonly used for determining radioactivity.

100  In the study of a biochemical pathway using a $^3H$ or $^{14}C$ labelled metabolite, is the position of the label in the molecule of importance?

# PROTEINS, STRUCTURE AND FUNCTION

## A   Sequencing of peptide chains

101   List methods that have proved to be of value in purifying proteins.

102   If a protein possesses more than one polypeptide chain, give 2 examples of chemical bonds by which the chains might be joined and show how the polypeptide chains could be separated in both cases.

103   Which methods are suitable for the separation of mixtures of amino acids obtained from protein hydrolysis?

104   Describe briefly the use of fluorodinitrobenzene in sequencing.

105   Name a reagent that may be used as an alternative to fluorodinitrobenzene in sequencing.

106   What are aminopeptidases and carboxypeptidases?

107   Which type of peptide bonds are readily hydrolysed by the enzymes trypsin and chymotrypsin?

108   Give an example of a chemical reagent that can produce specific cleavage of a polypeptide chain.

109   What is the Edman degradation and what is its value in sequencing?

110   Indicate how the overall amino acid sequence in a polypeptide may be elucidated.

## B   Conformation of peptide chains

111   What type of enzyme is required to catalyse the formation of the correct 3-dimensional protein configuration during or after synthesis of the polypeptide chain?

112   Give the main types of interaction that produce association between peptide chains or between different locations in the same chain.

113 The enormous range of 3-dimensional structural possibilities for polypeptides is greatly reduced by rigidity in the molecules of amino acids. Describe:
   (a) a molecular feature of all 20 amino acids that restricts the range of spatial conformations,
   (b) a molecular feature present in only one of the 20 amino acids that also restricts the conformation of polypeptides.

114 Describe how a polypeptide chain can form a stable helical structure ($\alpha$-helix).

115 Describe how polypeptide chains can associate to form sheets ($\beta$-pleated sheets).

116 Define what is meant by the terms primary, secondary, tertiary and quaternary structure of a protein.

117 Comment on whether the denaturation of a protein is always irreversible.

118 Give physical methods that have been used to investigate the 3-dimensional structures of proteins.

## C Types of protein

119 Traditional protein classification gave rise to names still in common usage. Two such names are albumin and globulin. How do these 2 classes differ in solubility?

120 What are scleroproteins?

121 What are histones?

122 Define the terms glycoprotein, lipoprotein and nucleoprotein.

123 Give an example of a metalloprotein containing:
   (a) iron,
   (b) zinc,
   (c) copper.

124 Give 3 examples of chromoproteins.

125 Describe what is meant by:
   (a) a phosphoprotein,
   (b) a flavoprotein.

13

126 Give 2 examples of groups of proteins which contain invariant sequences of amino acids together with other sequences that, if varied, do not remove the general class of biological activity that characterizes the whole group.

# OXYGEN-TRANSPORTING PROTEINS, HAEM AND PORPHYRINS

## A Haem and haemoproteins

127 Describe the structure of a haem group.

128 How many covalent-coordination valencies are possible for $Fe^{2+}$ or $Fe^{3+}$ in the haem structure and how are these arranged?

129 What is a haemoprotein?

130 Give examples of haemoproteins.

131 How does the strength of the association between haem and the polypeptide chain in haemoglobin compare with that in cytochromes?

132 How would one treat myoglobin to dissociate the haem from the globin?

133 How does the oxidation–reduction of the haem in haemoglobin differ from that in cytochromes?

134 Can the reaction $Fe^{2+} \rightarrow Fe^{3+} + e^-$ ever occur in haemoglobin?

135 Is there a turnover of haemoglobin in a particular red cell during the life span of this mature erythrocyte in circulation?

136 What is the approximate average daily turnover of haemoglobin in a normal adult male?

137 Is haem synthesized in body tissues other than bone marow and, if so, for what purpose?

## B Porphyrins and porphyria

[In questions and answers 138 to 151, the word porphyrin is abbreviated to (pn)].

138 Describe the structure of proto(pn)III (often called proto(pn)IX) and indicate its chief function.

139 Outline the biosynthesis of porphobilinogen (PBG), indicating where this occurs in the cell.

140 How are uro(pn)ogen I and III formed in the cytosol? What is the structural relationship between them? Which is known to have the more important function in the cell?

141 Indicate how the molecules of uro(pn)ogen, copro(pn)ogen, proto(pn)ogen, uro(pn), copro(pn) and proto(pn) are related. It is not necessary to give the structures of all these compounds.

142 Which of the compounds mentioned in question 141 are highly coloured?

143 By which route(s) may (pn)s and their precursors be excreted? What is the normal pattern of excretion?

144 What is the general biochemical basis for clinical conditions called porphyrias?

145 What symptoms appear to be clinically related to overproduction and high plasma levels of aminolaevulinic acid (ALA) and PBG?

146 Describe the symptoms that may be ascribed to high circulating levels of (pn)ogens and (pn)s.

147 Hereditary hepatic porphyrias are well known to possess latent and acute phases. What circumstances are known to precipitate an acute phase?

148 Describe briefly the different forms of inherited hepatic porphyrias. Indicate in each case, the defective stage in the haem synthetic pathway.

149 What are erythropoietic porphyrias?

150 Are there any conditions, classified as porphyrias, that are not primarily inborn errors of metabolism?

151 Are porphyrias the only conditions in which disorded porphyrin metabolism occurs?

## C  Myoglobin and haemoglobin

152  What are the main features of the polypeptide part of the complete 3-dimensional myoglobin structure?

153  Describe where the haem group is located in the globular protein and where the oxygen molecule attaches.

154  Describe the structure of the principal haemoglobin of normal adults' HbA.

155  What are the main functions of haemoglobin?

156  If the fraction of total oxygen-binding sites actually occupied is plotted against the oxygen partial pressure, a sigmoid curve is obtained. What does this indicate?

157  How do increases in $[H^+]$ and $[CO_2]$ affect the oxygen affinity of haemoglobin?

158  What is the effect of 2,3-diphosphoglycerate (DPG) on the oxygen affinity of haemoglobin?

159  The DPG content of erythrocytes is sensitive to changes in arterial oxygen partial pressure. Illustrate the value of this sensitivity.

160  Summarize the outstanding functional differences between myoglobin and haemoglobin.

161  How is the tetramer of fetal haemoglobin (HbF) related to that of adult haemoglobin (HbA)? What is the main functional difference between HbF and HbA?

## D  Haemoglobinopathies

162  Polypeptide chains, present in various normal human haemoglobins, are given as symbols the letters $\alpha$, $\beta$, $\gamma$, $\delta$, $\varepsilon$ and $\zeta$ of the greek alphabet. What types of normal human haemoglobins contain these various chains?

163  Do the $\alpha$-, $\beta$-, $\gamma$- and $\delta$-chains all contain the same number of amino acid residues?

164  Can fetal haemoglobin (HbF) production persist at a high level after the neonatal period?

165 In each case, give an example of an abnormal haemoglobin of clinical significance in which:
(a) a single amino acid substitution has occurred on the surface of the haemoglobin molecule,
(b) a single amino acid substitution has occurred near the active site of oxygen uptake.

166 Apart from the types of amino acid substitution in the previous question, in what other ways may haemoglobin abnormalities arise?

167 What is the usual way in which an abnormal HbA chain differs from an $\alpha$- or $\beta$-chain and do such abnormal haemoglobins always produce clinical effects?

168 What is the amino acid substitution in HbS?

169 Describe and comment on the differences in aqueous solubility between HbS, HbA, and their oxygenated derivatives.

170 Is any relation known between the frequency of the sickle cell gene and the incidence of infectious diseases in populations?

171 What are thalassaemias?

# ENZYMES

## A Classification and general properties

172 Give an illustrative example of each of the following classes of enzyme: oxidoreductases, transferases, hydrolases.

173 What are the 3 remaining enzyme classes? Illustrate each with an example.

174 Have enzymes ever been isolated as crystalline proteins?

175 Are all enzymes simple proteins?

176 Distinguish between the transient state, steady state and equilibrium state for a sequence of reactions.

177 Define the following terms:
(a) the velocity of an enzyme-catalysed reaction,
(b) the enzyme activity of an enzyme preparation,
(c) the specific activity of an enzyme.

178 What is meant by the order of a reaction? Distinguish between zero order, first order, and second order reactions.

179 Describe, in general terms, how a reaction that is immeasurably slow, in the absence of an enzyme, may become a rapid process if the appropriate enzyme is present.

180 Describe how the catalytic activity of an enzyme changes with increase in temperature.

181 Discuss briefly the dependence of enzyme activity on pH.

182 How is it possible for an optically inactive (i.e. symmetrical) molecule to be converted by an enzyme-catalysed reaction into a product that is a single optically active enantiomer?

## B Mechanism of action

183 What is the Michaelis–Menten mechanism for an enzyme-catalysed single substrate reaction?

184 How is the reaction velocity $v$ related to the substrate concentration [A], under steady state conditions, for a reaction obeying Michaelis–Menten kinetics?

185 What is the order of reaction of such a reaction (a) at very low [A] values and (b) at very high [A] values?

186 What type of mathematical curve is obtained when $v$ is plotted (in rectangular cartesian coordinates) against [A] for a Michaelis–Menten reaction?

187 A particular substrate concentration is identical to the Michaelis constant for the enzyme-catalysed reaction. What concentration is this?

188 What is a Lineweaver–Burk plot?

189 Some 2-substrate reactions are said to obey a 'ping-pong' mechanism. Describe such a mechanism.

190 The dehydrogenation of lactate to pyruvate is a 2-substrate reaction, described as possessing an ordered mechanism. Outline the mechanism and say why it is described as 'ordered'.

191 How may a particular 2-substrate reaction be identified as 'ping-pong' or 'ordered'?

192 Select a suitable proteolytic enzyme and indicate the nature of the catalytic process that occurs at its active site.

## C  'Classical' inhibition

193 Distinguish between a reversible and an irreversible enzyme inhibitor.

194 What is a competitive inhibitor?

195 What is a non-competitive inhibitor?

196 What is an uncompetitive inhibitor?

197 For each of the above 3 distinctive classes of reversible inhibitor, indicate the necessary biochemical reactions that occur in the case of an enzyme, a single substrate, and a single inhibitor.

198 Does an enzyme-catalysed reaction, inhibited in any of the above ways, obey Michaelis–Menten kinetics?

199 Describe the Lineweaver–Burk plots for reactions in the presence and in the absence of these inhibitors.

200 Define the term inhibitor constant $K_I$, this term referring to an inhibitor acting on a particular enzyme-catalysed, single substrate, reaction under specified conditions. How may $K_I$ be determined?

201 Do the classes of inhibitors discussed in the preceding questions apply to reactions with more than one substrate?

202 Why is it that administration of ethanol can be an effective treatment of poisoning by other alcohols, such as methanol or ethylene glycol?

## D Allostericity

203 Initial velocity measurements on an enzyme-catalysed reaction are plotted against substrate concentration. A curve is obtained in which the slope $(dv/d[A])$ is small near the origin, but rapidly increases when a particular (low) substrate concentration is reached so that a relatively small increase in substrate concentration produces a large increase in velocity. At higher substrate concentrations, the velocity levels out to approach asymptotically a maximum velocity. Is this a Michaelis–Menten type velocity and, if not, what type of mechanism would account for this behaviour?

204 The phenomenon described in the preceding question is usually exhibited in a particular type of enzyme structure. Which type?

205 What is the difference between positive and negative cooperativity?

206 How is it that a molecule present at a regulatory or allosteric site can influence the activity at a distant catalytic site in the enzyme molecule?

207 Catalytic activity can be 'switched on' at a suitably high substrate concentration. Are there cases in which the activity is 'switched off' by high substrate concentration?

208 Do allosteric effectors, either positive or negative, have to be substrate molecules?

209 If allosteric effectors are not substrates or products of the reaction, what is their physiological significance?

210 Describe what is meant by the term 'key enzyme'.

211 Give an example of allosteric inhibition acting as a negative feedback mechanism.

212 Give an example of allosteric activation producing a positive feedforward effect.

# E  Clinical enzymology

213 Give examples of isoenzymes that
(a) are present in different parts of a cell,
(b) occur in different cells,
(c) are present in different proportions in different tissues.

214 What general methods enable different isoenzymes to be distinguished and assayed for clinical purposes?

215 How may kinetic studies on a specimen give assay results for 2 isoenzymes?

216 Why are enzyme activities usually determined in serum rather than in plasma?

217 Quote 3 serum enzymes frequently investigated in relation to myocardial infarction. Give 2 circumstances in which knowledge of enzyme activities can be of special value in this disease.

218 Comment briefly on the value of creatine kinase assays in skeletal muscle disease.

219 Acid phosphatase activities are measured in blood in cases of carcinoma of the prostate. Explain why it is desirable that either unhaemolysed blood is used or measurements are made with both alcohol-treated and untreated samples. What is the value of the determinations?

220 Give examples of conditions with increased osteoblastic activity when serum alkaline phosphatase activity is commonly increased.

221 Why is it that hepatocellular damage may result in lower serum levels of some enzymes and raised serum levels of others?

222 Can serum enzyme determination give early indication of liver cell damage resulting from drug-taking over long periods?

223 What serum enzyme determinations are valuable in investigating (a) cholestasis and (b) acute pancreatitis?

# SOME IMPORTANT TYPES OF SUBSTRATE MOLECULE

## A Carbohydrates

224 Distinguish between an aldose and a ketose.

225 What structures could be considered as the simplest carbohydrates?

226 Define and give 2 examples of monosaccharides.

227 Distinguish between a pentose and a hexose, giving an example of each (name and structure).

228 Distinguish between a pyranose and a furanose structure and give an example illustrative of each.

229 Define the terms disaccharide and polysaccharide. Give an example of each, indicating the composition of the chosen disaccharide.

230 What is the relationship between a D- and an L-carbohydrate?

231 What is the structural difference between $\alpha$-D-glucose and $\beta$-D-glucose? Are these substances readily interconvertable?

232 How does a reducing sugar differ in structure from a non-reducing sugar?

233 Classify the following carbohydrates into broad classes: (a) amylose, (b) arabinose, (c) lactose, (d) fructose, (e) glycogen, (f) mannose.

## B Triglycerides

234 Of which alcohol are triglycerides the esters?

235 Are the biochemically important triglycerides lipophilic or hydrophilic molecules?

236 How may triglycerides be hydrolysed and what types of product are obtained?

237 What type of carbon chain is present in (a) stearic acid and (b) oleic acid?

238 Comment on the relative importance of the straight chain molecules $C_{15}H_{31}.COOH$ and $C_{16}H_{33}.COOH$ in human metabolic processes.

239 What general classification term is used to describe linoleic acid, arachidonic acid and other chemically related acids?

240 How is it that the double bonds in arachidonic acid can be described as $\Delta$ 5, 8, 11, 14 and as $\omega$ 6, 9, 12, 15?

241 Why is it that the $\omega$ system of nomenclature is more convenient than the $\Delta$ system in biochemistry?

242 Which molecule has the greater quantity of metabolically available energy (for humans), a carbohydrate with 12 carbon atoms or a saturated fatty acid with 12 carbon atoms?

## C Nucleotides

243 Distinguish between a nucleoside and a nucleotide.

244 Are nucleoside phosphates always esters of orthophosphoric acid?

245 Describe the differences between the pyrimidine and the purine ring systems.

246 Give the names and single letter abbreviations of the following bases and state which bases are pyrimidines and which are purines:

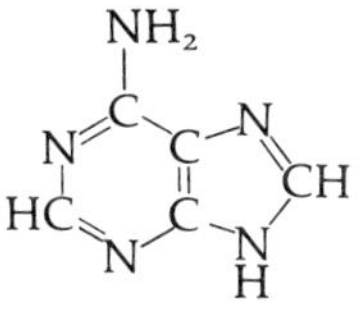

(a)

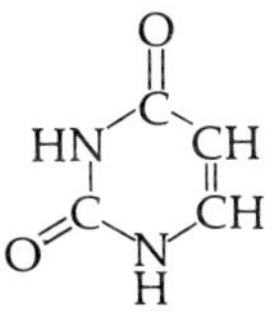

(b)

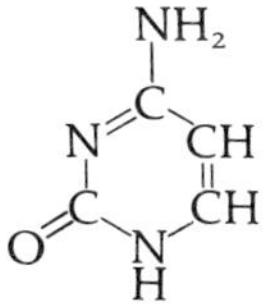

(c)

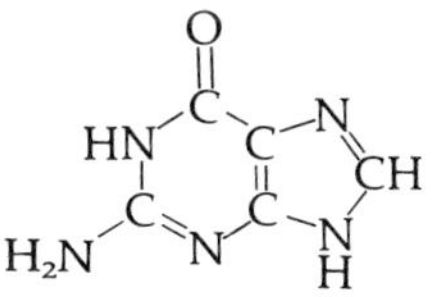

(d)

247 Name a pyrimidine base of outstanding significance not among structures (a) to (d) in the
preceding question. What is its relationship
chemically to one of the structures (a) to (d)?

248 Are the single letter abbreviations (A, U, C, G
and T) used to refer solely to the pyrimidine and
purine bases themselves?

249 Describe what is meant by a $\beta$-$N$-ribosidic linkage.

250 What type of product might be expected on mild
hydrolysis of a nucleotide (base–sugar–
phosphate) with (a) 2M hydrochloric acid and
(b) 2M sodium hydroxide?

251 Give formulae that show how dAMP differs from
AMP.

252 Will the ATP molecule in solution of pH 7 carry
an electric charge and, if so, what type of charge
is likely?

## D  Other significant substrate molecules

253 What is the molecular difference between NADH
and NADPH and what electrical charges, if any,
are carried by these molecules at pH 7?

254 How do the $NAD^+$ and $NADP^+$ structures alter
when these molecules act as hydrogen acceptors?

255 Give 3 illustrative examples, other than nucleotide derivatives, of high energy phosphoric acid
derivatives.

256 Give the main reason why the ATP structure has
great phosphorylating potential.

257 Describe the main features of the coenzyme A
molecule.

258 Comment on the nature and function of the prosthetic group in flavoproteins.

259 Briefly indicate the structural features and function of coenzyme $Q$ (ubiquinone) structures.
What is their structural similarity to vitamin K?

260 Another coenzyme of major importance is pyridoxal phosphate. Describe the key features of its chemical structure and show its chemical relationship to pyridoxamine phosphate.

# GLYCOLYSIS AND FATTY ACID OXIDATION

## A Glycolysis

261 Distinguish between aerobic and anaerobic glycolysis.

262 Identify the molecule that acts as hydrogen acceptor in aerobic glycolysis and explain the role of this substance in anaerobic glycolysis.

263 Which reaction in the glycolytic pathway involves the breaking of a C–C bond and what is the name of the enzyme involved?

264 Which reactions directly produce or remove ATP in anaerobic glycolysis and what is the net production of ATP per glucose molecule in the process?

265 In which part of the cell does glycolysis occur?

266 Give 2 examples where anaerobic glycolysis is important in human cells.

267 If the oxidation of an NADH molecule gives rise to 3 ATP molecules, what is the maximum overall production of ATP from the conversion of one glucose molecule into 2 pyruvic acid molecules?

268 Which steps in the glycolytic pathway can only, in effect, be reversed if the cell employs a different reaction mechanism with different enzymes?

269 Distinguish between glucokinase and hexokinase.

270 Give an example that demonstrates the important part glycolysis plays in the metabolism of hexoses other than glucose.

## B Fatty acid oxidation

271 Is ATP required in the oxidation of fatty acids?

272 In which region(s) of the cell does the oxidation of fatty acyl CoA derivatives occur?

273 What is carnitine and what part does it play in fatty acid oxidation?

274 Why is fatty acid oxidation often called $\beta$-oxidation?

275 Which are the coenzymes involved in the oxidation of fatty acyl CoA molecules?

276 In a cellular region where the oxidation of palmitoyl CoA $[C_{15}H_{31}.COS(CoA)]$ is proceeding, what other fatty acyl CoA derivatives are likely to be present?

277 What is the thiol ester that is the product of the complete fatty acid oxidation of stearic acid $(C_{17}H_{35}.COOH)$?

278 How many coenzyme A molecules are required for the complete $\beta$-oxidation of the straight chain fatty acid $C_{2n+1}H_{4n+3}.COOH$ (where $n$ is an integer)?

279 How many FAD and $NAD^+$ molecules are reduced in the complete $\beta$-oxidation of a molecule of palmitic acid $(C_{15}H_{31}.COOH)$ and how many ATP molecules could result from the passage of electrons from these $FADH_2$ and NADH molecules to oxygen?

280 Which phenyl-substituted carboxylic acid will be formed in the liver from the complete $\beta$-oxidation of 6-phenylhexanoic acid?

# CITRIC ACID CYCLE, ELECTRON TRANSPORT AND OXIDATIVE PHOSPHORYLATION

## A Citric acid cycle

281 Where in the cell does the oxidation of pyruvate occur?

282 Give a major consequence in mammals of the irreversibility of the transformation from pyruvate to acetyl CoA, catalysed by the pyruvate dehydrogenase enzyme complex.

283 What parts do thiamin pyrophosphate (TPP) and lipoic acid play in the conversion of pyruvate to acetyl CoA?

284 What process within the tricarboxylic acid cycle itself is analogous to the formation of acetyl CoA from pyruvate?

285 How does acetyl CoA (produced from carbohydrate and from fatty acids) enter the tricarboxylic acid cycle?

286 Which steps within the cycle produce $CO_2$?

287 Which reaction in the cycle converts guanosine diphosphate into guanosine triphosphate?

288 Which sequence of three reactions in the cycle is analogous to a sequence of three steps in the $\beta$-oxidation of a fatty acid?

289 Taking an NADH molecule as equivalent to 3 ATP and an $FADH_2$ molecule as equivalent to 2 ATP, what is the total ATP + GTP production available from the oxidation of a pyruvic acid molecule?

290 Mention other functions that the tricarboxylic acid cycle might possess in addition to that of catalysing the extraction of energy from acetyl groups.

## B  Electron transport and oxidative phosphorylation

291 Where in the cell do electron transport and oxidative phosphorylation occur?

292 Summarize the reaction sequence that occurs when an NADH molecule is oxidized and the electrons transferred to oxygen *via* the electron transport chain.

293 At which points in the above sequence are ATP molecules produced in the coupled process?

294 Outline the reaction sequence when $FADH_2$ from succinate is oxidized and the electrons transferred to oxygen by means of the electron transport chain. How many ATP molecules are produced in this sequence?

295 Where in the electron transport chain do cyanide and sulphide ions exert their blocking effects?

296 Is it possible for electron transport to proceed without concomitant ATP production?

297 What normally exercises control over the operation of the electron transport chain in living cells?

298 Give some examples of uncoupling agents.

299 In the past, some patients have been treated with uncoupling agents: for what purpose?

300 Indicate briefly the nature of the chemiosmotic hypothesis of Mitchell.


# THE FED STATE

## A Glycogenesis

301 Glycogen is a polysaccharide made entirely of $\alpha$-D-glucose residues. What is the nature of the majority of the linkages between the glucose residues?

302 Some linkages between glucose residues produce a branch in the chain of glucose residues. What are these linkages called and what is the molecular nature of the branch?

303 Do tissues other than liver and muscle contain glycogen?

304 In which region of the cell is glycogen synthesized and stored?

305 Is glycogen transported between tissues in the body?

306 How does the pathway of glycogen synthesis in liver differ from that in muscle?

307 Indicate briefly the biosynthetic route that produces unbranched $\alpha$-1,4-chains from glucose 6-
phosphate.

308 What biosynthetic process produces branching in
the chains?

309 Which are the hormones mainly concerned in
controlling the action of glycogen synthetase?

310 Name, with brief comments, an inborn error of
glycogen synthesis.

## B Lipogenesis

311 In which cell compartment does fatty acid synthesis occur?

312 What is the fatty acid synthetase complex?

313 Acetyl CoA, the starting point for fatty acid
synthesis, is formed from carbohydrate (or from
fatty acid) in the mitochondrion. How do the
acetyl groups get transferred to the cytosol for
fatty acid synthesis?

314 Describe briefly the process by which an acetoacetyl protein complex is formed from acetyl
CoA.

315 How is the acetoacetyl group reduced to the
butyryl group for fatty acid synthesis?

316 How is further chain elongation effected to give
palmitic acid?

317 What is the origin of $\alpha$-glycerophosphate in
adipose tissue or in intestinal mucosa?

318 Can $\alpha$-glycerophosphate be produced directly
from glycerol in human tissues?

319 How are triglycerides formed and which tissues
are the main producers?

320 Can mammalian cells further extend the carbon
chain of palmitic acid to give stearic acid and
even longer chain acids?

## C The pentose phosphate pathway

321 In which part of the cell is the pentose phosphate pathway (hexose monophosphate shunt) located?

322 In which tissues is the pentose phosphate pathway of particular importance?

323 Summarize the important functions of this pathway.

324 Give the reactions in this pathway that lead to the formation of a $CO_2$ molecule from glucose 6-phosphate.

325 What type of reactions in the pathway are catalysed by transketolases and what cofactors are necessary?

## D Unsaturated fatty acids

326 How many C=C double bonds are present in each molecule of oleic, linoleic, linolenic, and arachidonic acids? Which of these double bonds are *cis* and which *trans* in the naturally occurring isomers? Are *cis* and *trans* forms equally effective as human essential unsaturated fatty acids?

327 How can the oleic acid structure be formed biochemically in the body?

328 Can linoleic, linolenic, and arachidonic acids be synthesized in the human from saturated acids?

329 Which C=C double bonds of unsaturated fatty acids cannot be formed in mammalian cells?

330 Give examples of ways in which essential fatty acid deficiency might arise.

## E The hypoglycaemic hormone

331 Which substance or substances can stimulate pancreatic $\beta$-cells to secrete insulin?

332 Is insulin required for the uptake of glucose into all cells?

333 How does insulin exert its stimulating action on glycolysis in liver?

334 In addition to its effect on glycolysis, what other effects are exerted by insulin on carbohydrate and fat metabolism?

335 Give any effects that insulin may exert other than those on carbohydrate and fat metabolism.

# THE FASTING STATE

## A Fasting state hormones

336 Name important hormones that are concerned in controlling metabolism in the fasting state and the glands that secrete these hormones.

337 Indicate the types of molecular structures possessed by these hyperglycaemic hormones.

338 What is the outstanding function of all these hormones in relation to homoeostasis in the fasting state?

339 In what ways are the effects of adrenaline and glucagon similar in the fasting state?

340 What are the relevant activities of glucocorticoids in the fasting state?

341 Comment briefly on the anti-insulin activities of GH and ACTH.

## B Glycogenolysis

342 In which cell compartment of which tissues does glycogen breakdown occur?

343 Summarize the overall mechanism for the breakdown of unbranched portions of the glycogen molecule.

344 How are the chain branches degraded?

345 How is the glycogen catabolism controlled?

346 What is the main function of glycogenolysis in liver?

347 Can the considerable quantity of glycogen in skeletal muscle be used to maintain blood glucose level?

348 Which is the most common of the glycogen storage diseases and which enzyme is deficient in this condition?

349 Summarize the main features of the disease concerned in the previous question.

## C Gluconeogenesis

350 Summarize the purpose of gluconeogenesis.

351 In which tissues is gluconeogenesis particularly active?

352 In which of the cell compartments does gluconeogenesis occur?

353 Explain how removal of blood glucose by glycolysis in skeletal muscle can be offset by gluconeogenesis in liver.

354 Outline briefly how pyruvate for gluconeogenesis can be transferred from muscle in the form of an amino acid.

355 How are oxo acids converted into phosphoenolpyruvate?

356 Which other reactions of the gluconeogenesis pathway differ from (reverse) glycolysis reactions?

357 Comment briefly on the control of gluconeogenesis.

## D Lipolysis and ketogenesis

358 Where in the body does lipolysis occur?

359 Comment on the meaning of the term adipose tissue lipase.

360 What other enzymes catalyse the hydrolysis of free or bound triglyceride?

361 What happens to the glycerol produced by the action of adipose tissue lipase?

362 What happens to the fatty acids liberated by this lipase?

363 What type of control is exerted on adipose tissue lipolysis?

364 What are ketone bodies?

365 Where is acetoacetate synthesized?

366 What is the route by which ketone bodies are formed?

367 What function do ketone bodies possess?

## E   Diabetes mellitus

368 Summarize the common clinical characteristics of juvenile onset diabetes.

369 Which organ is functionally defective in juvenile onset diabetes?

370 Compare the results of glucose tolerance tests on normals and on patients with diabetes mellitus.

371 Are blood lipid concentrations modified in diabetes mellitus?

372 One of the complications of juvenile onset diabetes is the development of an acidosis. Discuss this briefly.

373 Give briefly clinical features of adult (mature) onset diabetes.

374 Is insulin formation impaired in adult onset diabetes?

375 Are diabetes mellitus and diabetes insipidus variants of the same disease? Summarize the cause and features of the latter condition.

# AMINO ACID METABOLISM

## A   The amino acid pool

376 What is considered to constitute the amino acid pool of the body?

377 Indicate the types of nitrogen-containing compounds that may be synthesized from amino acids sequestered from the pool.

378 For what other major purpose may amino acids be required from the pool?

379 Does the amino acid pool constitute a large storage of amino acids (comparable in size to fat storage depots)?

## B Formation of oxo acids from amino acids

380 Describe the reaction catalysed by glutamate dehydrogenase.

381 Give the site of action of glutamate dehydrogenase and comment on the significance of this enzyme in metabolism.

382 What type of reaction is catalysed by L-amino acid oxidase and where is this enzyme located?

383 Which human tissues contain D-amino acid oxidase and what metabolic function does this enzyme possess?

384 Give examples of non-oxidative deamination reactions.

385 Give examples of some nitrogen-containing substances that can be produced metabolically from $NH_3$ or $NH_4^+$.

386 What is transamination?

387 In which cell compartments of which tissues do transamination reactions occur?

## C Urea cycle and nitrogen balance

388 What is the purpose of the urea cycle?

389 Which part of the cell contains urea cycle enzymes?

390 Which are the tissues in which the urea cycle operates?

391 How does combined nitrogen enter the cycle?

392 How does nitrogen leave the cycle?

393 Name the α-amino acids that are components of the cycle. Which of these can be incorporated into protein structure?

394 Is ATP required as a cosubstrate for any step(s) in the cycle and, if so, for which step(s)?

395 If an averagely active healthy man consumes 100 g of protein a day (virtually his total combined nitrogen intake), approximately how much nitrogen should he excrete daily and how is this excreted?

396 Give examples of circumstances when an individual might have (a) a positive nitrogen balance and (b) a negative nitrogen balance.

397 What is kwashiorkor?

## D Decarboxylation of amino acids

398 Give examples of amino acids that undergo decarboxylation.

399 What type of compound is produced by decarboxylation of amino acids?

400 Give examples of amines produced either directly or indirectly by decarboxylation of amino acids.

401 What cosubstrate is required in the decarboxylation reactions?

402 In which cell compartment does amino acid decarboxylation occur? Comment on the tissues involved.

403 Give the biosynthetic pathway for noradrenaline and adrenaline.

404 Where does the above pathway operate?

405 What is melatonin?

## E Inborn errors of amino acid metabolism

406 Indicate briefly how inborn errors arise.

407 List some inborn errors of phenylalanine catabolism.

408 Describe how phenylalanine is hydroxylated to give tyrosine.

409 Which enzyme is defective in phenylketonuria?

410 What consequences follow from this enzyme defect?

411 What steps are taken to detect and to treat phenylketonuria?

412 What is alkaptonuria?

413 The characteristic odour of the urine of a patient with maple syrup urine disease is probably due to the presence of low molecular weight, branched chain, aliphatic $\alpha$-hydroxy acids. What would be the origin of such compounds in the urine?

414 Why is the term cystinuria not an ideal name for the referred to condition?

415 What enzyme is deficient in homocystinuria and why are connective tissue abnormalities found in this condition?

416 What is a common consequence of deficiencies in any of the urea cycle enzymes?

# NUCLEOTIDE METABOLISM

## A Biosynthesis of IMP, AMP and GMP

417 What is the unusual aspect of the kinase-catalysed formation of 5-phosphoribosyl 1-pyrophosphate (PRPP) from ribose 5-phosphate?

418 Outline briefly the origin of the various (numbered) atoms or groups in the synthesis of IMP (inosine monophosphate):

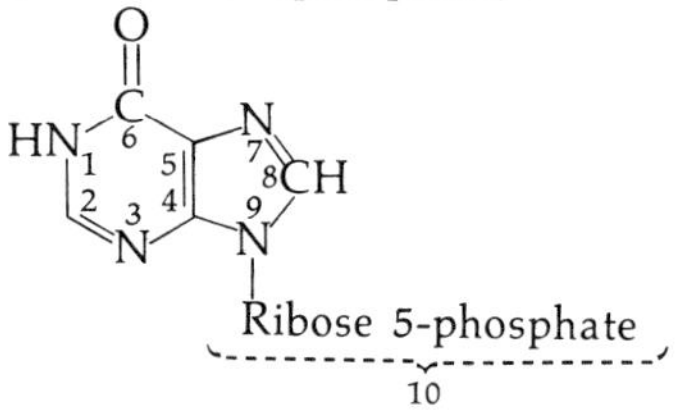

(IMP)

419 Comment on the structure and significance of the tetrahydrofolic acid derivatives that are involved in purine synthesis.

36

420 What is meant by a folic acid antagonist? Give an example of such a compound.

421 Name some other antineoplastic agents that inhibit either purine biosynthesis or the interconversion of purines.

422 How is AMP formed from IMP?

423 How is GMP formed from IMP?

## B Biosynthesis of UMP and of ribonucleoside triphosphates

424 What is the outstanding difference in the roles played by PRPP in purine and in pyrimidine synthesis?

425 In pyrimidine synthesis, 4 atoms of the ring come from one amino acid. Which amino acid?

426 How is the pyrimidine ring structure formed?

427 How is orotate converted into UMP?

428 Why do liver cells contain both mitochondrial carbamoyl phosphate synthetase and a different cytoplasmic carbamoyl phosphate synthetase with corresponding separate carbamoyl phosphate pools?

429 How are nucleoside triphosphates, other than ATP, formed in cells?

430 Indicate how CTP is formed from UTP.

## C Deoxyribonucleotides

431 Which molecules are the precursors of the purine and pyrimidine 2'-deoxyribonucleotides?

432 Is a formyl derivative of tetrahydrofolate ($FH_4$) required in any pyrimidine deoxyribotide synthesis?

433 What is the main site of purine and pyrimidine nucleotide synthesis?

434 How is control exercised over nucleotide biosynthesis?

435 List some artificial pyrimidine derivatives and analogues that act as antimetabolites.

## D Uric acid, gout and the Lesch–Nyhan syndrome

436 Indicate how pyrimidine and purine bases are produced by catabolism of nucleic acids.

437 How is uric acid produced in liver, kidney and other tissues?

438 What type of prosthetic group is possessed by xanthine oxidase?

439 What is allopurinol and what product is formed when this compound is oxidized in the presence of xanthine oxidase?

440 What are the features of gout?

441 What is the probable origin of acute inflammation in gout?

442 Indicate possible causes of gout.

443 What drugs are useful in treating gouty inflammation?

444 Describe the Lesch–Nyhan syndrome.

445 What is the biochemical basis of the Lesch–Nyhan syndrome?

# NUCLEIC ACIDS AND PROTEIN SYNTHESIS

## A DNA structure and replication

446 Describe the molecular structure of a single DNA chain.

447 What are the essential 3-dimensional features of double-standard DNA?

448 Indicate the conditions under which DNA polymerase I will act, *in vitro*, to produce a DNA-type polymer.

449 What is a DNA ligase?

450 What is circular DNA and where is it found?

451 Are deoxypolynucleotide chains synthesized in both $5' \rightarrow 3'$ and $3' \rightarrow 5'$ directions?

452 Summarize the mechanism that is accepted as plausible for DNA replication.

453 Briefly compare the properties of DNA polymerases II and III with those of DNA polymerase I.

454 What carries the genetic information in DNA?

455 What is meant by the term 'recombinant DNA'?

## B Biosynthesis of RNA

456 Name the major types of RNA found in cells.

457 How do RNA molecules differ from those of DNA?

458 What molecular components participate when mRNA is synthesized?

459 In which direction does new RNA synthesis extend an already formed RNA chain?

460 Where in the mammalian cell is RNA polymerase found?

461 Is the RNA biosynthetic process reversible?

462 In RNA biosynthesis, are both strands of double-stranded DNA transcribed and can single-stranded DNA acts as a template?

463 Describe the action of reverse transcriptase.

## C Activation of amino acids

464 What are the nucleotide residues that terminate tRNA chains?

465 Describe other general features that characterize tRNA structures.

466 Describe the process that attaches an amino acid to its appropriate tRNA molecule.

467 Specific sites exist in the tRNA molecule for attachment to other molecular structures. What are these structures?

468 Does a particular cell possess only a single tRNA molecule for each of the 20 amino acids?

## D  Coding and translation

469 What constitutes the genetic code for a particular amino acid?

470 Could pairs of adjacent bases (2-letter words) form codons?

471 Why is the code described as degenerate?

472 What are chain-terminating triplets and how many of these exist?

473 What signals the start of polypeptide chain synthesis?

474 Describe the formation of the 70S ribosomal initiation complex for protein biosynthesis in *E. coli.*

475 Describe briefly the ribosomal process that forms the first peptide bond.

476 What is meant by 'translocation'?

477 Comment on the chain-terminating process.

478 Summarize the directions along the chains in which the reading of the sequences occurs in transcription and in translation.

479 What is a mutation and what is meant by transition and transversion mutations and by frame-shift mutations?

480 What is a polyribosome (or polysome)?

481 List some post-ribosomal modifications of proteins.

482 How do certain cells produce secretory protein?

483  Give an example illustrating the mechanism by which synthesis of a particular protein may be regulated in a bacterial cell.

## E  Mechanism of action of some antibiotics

484  Comment briefly on the main effect of low concentrations of actinomycin D on cellular processes.

485  How may the activity of actinomycin D be exploited?

486  Name an inhibitor of transcription that has proved of notable therapeutic value and give its site of action at the molecular level.

487  Puromycin has been of great value in investigations on the mechanism of protein synthesis. What type of chemical structure does this antibiotic possess and how does the antibiotic interfere with translation?

488  How does streptomycin interfere with protein synthesis?

489  Describe how chloramphenicol and cycloheximide inhibit protein synthesis.

490  Give examples of other antibiotics that inhibit protein synthesis.

# LIPIDS AND STEROIDS

## A  Membrane lipids

491  What main classes of lipids are present in membranes?

492  Give examples of types of glycerophospholipids.

493  Give examples of biologically important derivatives of the long chain, unsaturated amino alcohol, sphingosine.

494  Indicate the importance of cytidine triphosphate (CTP), fatty acyl CoA and serine in the biosynthesis of these types of lipid structures.

495 Describe briefly the fluid mosaic model of membrane structure.

496 The free lateral movement of lipid membrane components is greatly influenced by the presence of uncharged (neutral) lipid. Comment on this statement.

497 Carbohydrate chains of glycoprotein or glycolipid lie at, or extend from, the outer surface of a cell membrane. What function do these structures possess?

498 By what processes are solute molecules transported across cell membranes?

499 Give some examples of inherited lipidoses, indicating general clinical features of these conditions. In one case name the defective enzyme and give the consequences of this enzyme defect.

500 What is the respiratory distress syndrome with hyaline membrane disease?

## B Cholesterol

501 Describe briefly the main features of the cholesterol molecule.

502 What type of tissue contains cholesterol?

503 Cholesterol can form esters with fatty acids. In which human tissues and by which reaction processes can these esters be formed?

504 Which organ is the main site of cholesterol synthesis and from which 2-carbon unit is the cholesterol molecule synthesized?

505 $\beta$-Hydroxy-$\beta$-methylglutaryl CoA is synthesized by the same reaction sequence in both mitochondria and cytosol. Outline this sequence and give the reason for synthesis in two subcellular sites.

506 What is the key step, at which regulatory factors operate, in the biosynthesis of squalene, the open chain precursor of cholesterol?

507 What are isoprenoid units and how are they formed?

508 Indicate how cholesterol is synthesized from iso-prenoid units.

509 Indicate the factors that control the synthesis of cholesterol in liver.

510 Comment briefly on the relation of cholesterol and its esters to atherosclerosis.

## C  Bile acids

511 What are the bile acids?

512 From which monohydroxysteroid are the primary bile acids synthesized and in which organ does this synthesis occur?

513 In what respect does the molecule of cholic acid differ from that of chenodeoxycholic acid? Disregarding details of stereochemical orientations, indicate briefly the biosynthetic pathway that produces the CoA derivatives of these acids.

514 What are the main bile acid components present in freshly secreted bile?

515 Where are secondary bile acids formed and what chemical relationship exists between the molecules of secondary and the molecules of primary bile acids?

516 Give two processes that exert control over the rate of biosynthesis of bile acids.

517 How is the formation of bile acids related to the elimination of cholesterol from the body?

518 How is it that freshly secreted bile contains glycine and taurine conjugates of secondary as well as of primary bile acids?

## D  Steroid hormones

519 Name the 5 main groups of steroid hormones that are derived metabolically from cholesterol and give, in each case, the major site of formation of these steroids.

520 Name an example of each of these 5 groups of steroid hormones and briefly indicate the physiological functions of the selected hormones.

521 In the conversion of cholesterol into the various steroid hormones, the common intermediate pregnenolone is formed. Outline briefly the reaction sequence that produces pregnenolone, showing clearly the chemical relationship of pregnenolone to cholesterol.

522 Which hormone stimulates the conversion of cholesterol to pregnenolone and where does this hormone originate?

523 Describe the molecular transformations that occur in the conversion of pregnenolone into progesterone.

524 What type of molecular change takes place when corticosteroids are formed from progesterone?

525 How is the aldosterone molecule related to that of corticosterone?

526 Outline the pathway by which progesterone is converted into other sex hormones.

527 What type of enzyme is involved in hydroxylations and certain other oxidations of steroids? Indicate, by a chemical equation, the type of overall reaction involved.

## E Plasma lipoproteins

528 What are the 4 major groups of plasma lipoproteins?

529 Give the order in which separation of these groups occurs in the ultracentrifuge.

530 Describe the separation that occurs on electrophoresis of (non-fasting) normal plasma lipoproteins at pH 8. Identify these globulin fractions in terms of the ultracentrifuge fractions.

531 Indicate the particle size and composition of the 4 plasma lipoprotein fractions.

532 Comment on the protein components of human plasma lipoprotein fractions.

533 Summarize briefly the process by which lipid compounds of dietary origin are transferred to adipose tissue, muscle and liver.

534 Summarize the process by which endogenous lipid compounds are transported from the liver to peripheral cells.

535 Where is HDL synthesized and what are its main roles?

536 List some inherited disorders concerned with lipoprotein metabolism and give a brief comment on each.

537 Will patients with xanthomatosis or with premature cardiovascular disease have abnormal plasma lipoprotein levels?

538 Give three disease conditions that often, as a secondary consequence, give rise to hyperlipoproteinaemia with raised plasma cholesterol and triglyceride concentrations.

# CONTROL OF METABOLISM

## A Allosteric control

539 List the main factors that control the velocity of a subcellular biochemical pathway.

540 What is meant by the term allosteric control?

541 What, if any, generalization may be made concerning the type of molecule that will act as an allosteric effector in any particular case?

542 Can an enzyme possess 2 (or more) allosteric sites that differ in their affinities for various allosteric effectors?

543 Are allosteric inhibition and activation major factors in the control of metabolism?

544 Give an example of negative feedback allosteric control.

545 Give an example of positive feedforward allosteric control.

546 Give an example of:
   (a) a substrate for a reaction that acts as an allosteric inhibitor for the same reaction;
   (b) a product of a reaction that acts as an allosteric activator for the same reaction.

## B Regulation by substrate supply and compartmentation

547   Explain and comment on how the availability of substrates, such as ATP, ADP, $NAD^+$, NADH, can have a very pronounced effect on the rates of metabolic processes, even if no allosteric effects are involved.

548   Show how substrate availability operates to limit the conversion of pyruvate to acetyl CoA when pyruvate is being formed from alanine in the liver.

549   The ratio of concentrations [$\beta$-hydroxybutyrate]/[acetoacetate] in a tissue can be taken as indicating the reducing (electron donating) state in the mitochondria. Suggest a ratio of concentrations that will reflect the cytoplasmic reducing state.

550   Bearing in mind that NADH, NADPH and similar molecules cannot traverse the mitochondrial membrane, indicate how hydrogen atoms (reducing power) are transferred between mitochondria and cytosol.

551   Describe how acetyl groups are transferred across the mitochondrial membrane.

## C Hormone control

552   Indicate the 2 mechanisms by which the 2 main types of hormone act to modify the enzyme activity of a target cell.

553   Give some examples of hormones of the first (quicker acting) type (in 552).

554   Indicate briefly how adrenaline stimulates glycogen phosphorylase.

555   Indicate briefly how adrenaline and ACTH stimulate, and insulin inhibits, the action of adipose tissue lipase.

556   What intracellular processes remove
    (a) cAMP,
    (b) phosphate ligands introduced into enzyme proteins by the action of cAMP?

557 What is a futile synthesis and show how this is avoided in the case of glycogenolysis?

558 List some processes that are known to be stimulated by cAMP.

559 List seven polypeptide or glycoprotein hormones synthesized in the adenohypophysis (of the pituitary) and indicate the function of each.

560 What type of molecular structure is possessed by the hypothalamic regulatory hormones? What is the regulatory function exerted by these hormones?

561 Name the hypothalamic hormone that stimulates, and the hypothalamic hormone that inhibits, the adenohypophyseal hormone somatotropin (growth hormone).

562 Summarize the hormonal regulation of cortisol production.

## MINERAL METABOLISM

### A Sodium, potassium and water

563 Name, for man, the cation and anion present in highest concentration in extracellular fluid and the metal cations present in the highest and the second highest concentrations in intracellular fluid.

564 Outline what is meant by the term 'sodium pump'.

565 What are the routes by which the body excretes $Na^+$ ions?

566 Describe where in the kidney tubules the processes of reabsorption and secretion of $Na^+$, $K^+$ and $Cl^-$ occur.

567 Outline the role that aldosterone plays in regulating the volume of the body's extracellular fluid.

568 In addition to the kidney, at what other sites does aldosterone influence the active transport and excretion of $Na^+$ ions?

569 What is the action of vasopressin (ADH) on the excretion of water?

570 Summarize the functions of $K^+$ ions in man.

571 Indicate the main causes of hyperkalaemia and suggest 2 possible approaches that might help in correcting this condition.

572 Indicate the main causes of hypokalaemia.

## B Calcium, magnesium and phosphorus

573 What are the approximate masses in g of calcium, magnesium and phosphorus in an average normal 70 kg man?

574 What are the main functions of calcium in the body?

575 What are the functions of magnesium in the body?

576 In what combined form is phosphorus present in the body?

577 Describe how calcium is carried in the plasma and comment on the magnitude and meaning of the plasma 'calcium' concentration.

578 What are the clinical effects of hypocalcaemia (low plasma $Ca^{2+}$ concentration) in man?

579 What are possible consequences of hypercalcaemia (high plasma $Ca^{2+}$ concentration) in man?

580 What are the 3 main materials known to exert physiologically controlling effects on the plasma $Ca^{2+}$ concentration?

581 Elaborate on the meaning of the term 'vitamin D'.

582 How does the body acquire vitamin D?

583 What is the active metabolite of vitamin D and how is this metabolite produced?

584 What substances control the 1-hydroxylase enzyme activity of kidney?

585 Briefly indicate the nature of the process by which 1,25-dihydroxycholecalciferol (1,25-DHCC) stimulates the intestinal uptake of $Ca^{2+}$ ions.

586 What are the main molecular features of the structures of parathyroid hormone (PTH) and calcitonin?

587 Summarize the physiological actions of PTH.

588 Summarize the physiological actions of calcitonin.

589 What type of biochemical mechanism would one expect to be involved in the actions of PTH and calcitonin?

590 Explain briefly the bone erosion that occurs in rickets.

591 If the plasma $Ca^{2+}$ concentration is normal, but the plasma phosphate concentration is low, what homoeostatic process can act to raise the plasma phosphate concentration?

## C Iron and other transition metal ions

592 Give an approximate value for the iron content of an adult male and indicate how this iron is distributed in various compartments.

593 How is iron lost from the body and how is this loss offset?

594 Give 3 possible reasons for the development of iron-deficiency anaemia.

595 Give examples of circumstances that can give rise to iron overload. Name 2 iron overload syndromes and indicate their probable causes.

596 How much copper is present in the adult human being? Give examples of its main functions.

597 Describe Wilson's disease.

598 Indicate the significance of the element zinc in human metabolism.

599 Indicate the part played by cobalt in human metabolic processes.

## D Anions (excluding phosphate)

600 Comment briefly on the relation of fluoride ions to health.

601 Give an example of maintenance of electrical neutrality by the passage of chloride ions through a cell membrane in the opposite direction to the passage of bicarbonate ions.

602 Give an example of migration of chloride ions when a cell retains bicarbonate ions and exports hydrogen ions.

603 What are rich sources of iodide ions in the human diet and how is iodide employed in the body?

604 Summarize briefly the physiological actions of the thyroid hormones.

605 Describe the usual process by which sulphate ions are converted into organic sulphates in the body.

# LIVER AND BILE

## A Functions of the liver and some inborn errors

606 Briefly summarize the general metabolic functions of the liver.

607 What other functions of biochemical interest are possessed by the liver?

608 Name two inborn errors of carbohydrate metabolism that involve liver damage and may present early in infancy as liver disease in an infant who is failing to thrive.

609 Outline the pathway of galactose metabolism in infants indicating the defect that causes glactosaemia.

610 What are the clinical features of galactosaemia and how is the condition treated?

611 What enzyme is defective in glycogen storage disease type I? What pathways will be influenced by this enzyme defect?

612 What are the clinical features of glycogen storage disease type I?

613 Comment briefly on other types of glycogen storage disease.

## B Bile pigment metabolism

614 What happens to the haemoglobin when senescent red cells are broken down in the spleen and elsewhere in the reticuloendothelial system?

615 What is the fate of newly formed bilirubin?

616 What happens to newly formed conjugated bilirubin?

617 What is the van der Berg reaction?

618 Are bilirubin, urobilinogen and urobilin present in normal human urine?

## C Jaundice

619 Give a definition of the term 'jaundice'.

620 Comment briefly on the classification of jaundice as prehepatic, posthepatic and hepatic.

621 Explain how jaundice, caused by unconjugated bilirubin, might develop.

622 Explain how the following might develop:
(a) posthepatic jaundice,
(b) hepatic jaundice.

623 What effects would almost total biliary obstruction have on urine composition in relation to:
(a) bilirubin,
(b) urobilinogen?

624 Why is jaundice common in the newborn?

625  Comment on the effects of drugs and toxins on the liver.

## D   Biochemical tests and liver disease

626  What is the BSP (bromosulphthalein) test? Indicate briefly how it is carried out and comment on its sensitivity.

627  What quantitative tests are used as an index of the synthesizing power of the liver? How is one such test adapted to help distinguish cholestasis from liver cell damage?

628  Give examples of enzymes that increase in concentration in plasma when there is liver cell damage, the concentrations being usable as a sensitive indicator of such damage.

629  Give examples of enzymes whose levels in plasma increase when cholestasis is present.

630  Can any information be gained from globulin electrophoretic patterns that is of value in relation to liver disease?

631  What biochemical test results are desirable when jaundice is first noticed in a patient?

632  Indicate any other types of biochemical (or immunological) tests of value in the diagnosis of liver disease.

## E   Bile

633  What volume of bile is produced each day by the liver and what types of component does it contain?

634  What is the function of the gall bladder in relation to the composition of the bile?

635  What functions are performed by the bile when it is expelled from the gall bladder?

636  Distinguish between cholesterol gall stones, pigment gall stones and mixed gall stones.

# BLOOD AND URINE

## A   The blood clotting process

637   Which vitamin is involved in the blood clotting process and what type of chemical structure does this vitamin possess?

638   Give the circumstances that can lead to deficiency of this vitamin in humans.

639   What is the effect on blood clotting of dicoumarol and warfarin?

640   Comment on the effects on blood clotting produced by (a) heparin, (b) citrate, (c) fluoride.

641   What is meant by the platelet response to blood vessel damage?

642   How is fibrinogen in blood converted into fibrin to form a clot? What is the name of the enzyme required for this process?

643   How is the fibrin matrix of the clot strengthened and made more rigid?

644   What happens to the clot as healing of the damaged blood vessel proceeds?

645   What is the protein precursor of thrombin called and what cofactors are required in the formation of thrombin from this precursor?

646   Comment on the term thromboplastic activity.

647   What is classical haemophilia or haemophilia A?

648   Indicate briefly the mode of action of vitamin K.

## B   Serum proteins

649   What are the main functions of serum proteins?

650   What separation process is used routinely to separate the main serum protein fractions and what types of technique have proved of value for the assay of special constituents?

651 Give brief explanatory comments on the following serum components:
(a) $\alpha_1$-antitrypsin,
(b) haptoglobin,
(c) complement.

652 Describe briefly the general structure of immunoglobulins.

653 What are the 5 main classes of immunoglobulins?

654 Which electrophoretic fractions contain the following proteins: haptoglobin, $\alpha_1$-antitrypsin, $\alpha_2$-macroglobulin, transferrin and IgG?

655 What underlying disturbances might be responsible for hypoalbuminaemia?

656 Describe the abnormalities in serum electrophoretic patterns often seen in:
(a) the nephrotic syndrome,
(b) cirrhosis,
(c) hypogammaglobulinaemia,
(d) monoclonal gammopathy.

## C The kidney and urine

657 Summarize briefly the functions of the glomerulus and of the proximal convoluted tubule of the nephron.

658 What factors can operate to reduce the glomerular filtration rate (GFR)?

659 Summarize briefly the functions of the loop of Henle, the distal convoluted tubule and the collecting duct.

660 Renal tubular dysfunction can often arise from damaging effects of components of the glomerular filtrate (GF). Give examples.

661 What is meant by the term 'threshold substance'?

662 Distinguish between overflow and renal aminoacidurias.

663 What is the Fanconi syndrome?

664 What type of disease is cystinuria?

665 Give the various types of renal calculi and an approximate guide to the proportion of cases of renal stone of each type.

666 Describe briefly the process in the proximal tubular cells that effects the recovery of bicarbonate ions from the GF.

667 Describe the processes that limit the acidity of the fluid passing through the distal tubule in spite of the secretion of hydrogen ions into the lumen.

668 What are the main components of normal urine?

669 What are the simple techniques that permit rapid testing for important abnormal components in fresh urine?


# BIOCHEMISTRY OF TISSUES

## A  Connective tissue (general)

670 Indicate the types of component that are present in connective tissue.

671 Describe the composition of a typical single polypeptide chain of collagen.

672 Describe the tropocollagen molecule.

673 What are collagen fibrils and how are they formed?

674 Summarize briefly the synthetic processes that produce a mature collagen fibre.

675 Does the collagen of normal tissue undergo catabolism?

676 Distinguish between proteoglycans and glycoproteins.

677 Describe, with 2 examples, the structure of glycosaminoglycan (mucopolysaccharide) chains.

678 Indicate an important function of the proteoglycan structure.

679 Indicate how rapidly connective tissue glycosaminoglycans are broken down and resynthesized in normal tissue.

680 What type of condition is included in the term mucopolysaccharidosis? Give 2 examples.

## B Mineralization and the composition of bone and teeth

681 What are the functions of osteoclasts, osteoblasts and osteocytes?

682 Describe briefly the extracellular components of bone.

683 Discuss briefly the composition and crystal structure of biological apatite minerals.

684 Comment briefly on similarities and differences in the composition of calcified cartilage, bone, dentine and cementum.

685 Outline briefly the composition of mature dental enamel.

686 Which anion is known to have a protective effect on dental enamel and how does it act?

687 Give the equation that defines the solubility product of the salt, calcium hydrogen phosphate. What relevance does this equation have to the solubility of biological apatite in serum?

688 Comment on the statement that 'plasma is supersaturated with respect to calcium hydroxyapatite'. If this statement is true, why does not mineralization take place in plasma?

## C Muscle and nerve

689 Describe the main structural features of a mature skeletal muscle cell.

690 What protein components of a sarcomere are directly concerned in the muscle contraction process and how are they arranged?

691 What type of molecular conformations are possessed by the main proteins concerned in voluntary muscle contraction?

692 Outline the mechanism of the contraction of striated muscle.

693 What is the role of $Ca^{2+}$ in muscular contraction?

694 Discuss briefly the source of energy for the contraction of striated muscle.

695 What is the role of acetylcholine in muscle contraction?

696 Give, with the chemical equation, the reaction catalysed by acetylcholinesterase. What happens as a consequence of the activity of this enzyme?

697 Give 2 examples of catecholamine neurotransmitters and indicate the reactions that inactivate these transmitter substances.

698 What type of neurotransmitter is $\gamma$-aminobutyrate and what is the reaction sequence that inactivates this transmitter?

# METABOLISM OF FOREIGN COMPOUNDS

## A General principles, absorption and excretion

699 Which of the 2 characteristics, lipid solubility or water solubility, will promote the absorption of foreign substances into the body, their diffusion into cells and their passive transport across membranes such as the blood–brain barrier?

700 Suggest a major route for the excretion of diethyl ether and other highly volatile materials.

701 Give examples of some types of foreign substances that may in practice be absorbed through the tissues of: (a) lung and (b) skin.

702 Why is it that weakly acidic drugs, taken by mouth, are absorbed mainly from the stomach rather than from the intestinal tract?

703 Where in the gastrointestinal tract is the absorption of weakly basic drugs likely to occur?

704 Excluding its antibacterial activity, what feature of the neomycin molecule makes it valuable for treating intestinal infections?

705 Will a weak acid (or a weak base) be more readily excreted if, in the kidney tubules and collecting ducts, it is present in the ionized or in the un-ionized form?

706 Does metabolic modification in the liver reduce the toxicity of all xenobiotics?

707 Give an example of metabolic modification: (a) decreasing and (b) increasing a pharmacological activity.

## B Phase 1 reactions

708 Reactions that metabolically modify drugs and other xenobiotics are sometimes classified as phase 1 and phase 2 reactions. What is the distinction between these?

709 Which subcellular fractions of liver cells contain enzymes that can effect the following changes in many appropriate xenobiotics: (a) oxidation, (b) reduction, (c) hydrolysis?

710 Indicate the phase 1 reactions involved in the processes of: (a) aromatic hydroxylation, (b) deamination and N-dealkylation.

## C Phase 2 reactions

711 What is the physiological effect of many types of phase 2 reactions?

712 Give examples of 2 amino acids that are conjugating agents in phase 2 reactions and give an example of this conjugation process.

713 Indicate the types of conjugation reaction that produce derivatives of D-glucuronic acid.

714 What type of molecule can undergo conjugation to form a sulphate?

715 Name some other types of conjugated derivatives.

# 3 Answers

## ACIDS AND BASES

### A  Proton transfer and pH

1  Lowry and Brønsted defined an acid as a molecule or ion that acts as a proton donor. Similarly a base was defined as a molecule or ion that acts as a proton acceptor. These are now the accepted definitions in biochemistry and medicine.
[E:94]

2  The equilibrium can be represented as

$$(HB)^{z+} + H_2O \rightleftharpoons H_3O^+ + B^{(z-1)+}$$

| conjugate | water | hydronium | conjugate |
| acid | | ion | base |

where $z$ is an integer $(0, \pm1, \pm2, \text{etc.})$. When a single acidic molecule or ion contains 2 ionizable hydrogen atoms, the above equation still applies, but 2 acids are present in the solution, the second acid being the conjugate base of the first acid.

3  The equation is

$$\frac{[H^+][\text{base}]}{[\text{acid}]} = K_a$$

where $K_a$ is a constant (at constant temperature), called the ionization constant of the acid.
[A:1]

4  A water molecule can transfer a proton from itself to another water molecule as follows:

$$H_2O + H_2O \rightleftharpoons H_3O^+ + OH^-$$

| proton | proton | hydronium | hydroxyl |
| acceptor | donor | ion | ion |

59

The amount of ionization is very small and so the $H_2O$ molecule can act both as a very weak base and as a very weak acid.
[H:10]

5   The equation is $[H^+][OH^-] = K_w$. The constant $K_w$ is called the ionic product for water. Its value is usually taken to be $10^{-14}$.
[E:93]

6   The classification is as follows: $H_2CO_3$ (acid), $CO_2$ (neither), $CO_3^{2-}$ (base), $CN^-$ (base), $NH_3$ (base), $NH_4^+$ (acid), $OH^-$ (base), $CH_3.CHOH.COO^-$ (lactate) (base), $CH_3.COOH$ (acid), $H_3O^+$ (acid), $H_3\overset{+}{N}.CH_2.COOH$ (acid).
[I:79]

7   These are capable of acting as acids or as bases, depending on the reaction, as indicated:

$$H_2CO_3 \underset{+H^+}{\overset{-H^+}{\rightleftharpoons}} HCO_3^- \underset{+H^+}{\overset{-H^+}{\rightleftharpoons}} CO_3^{2-}$$

$$H_3PO_4 \underset{+H^+}{\overset{-H^+}{\rightleftharpoons}} H_2PO_4^- \underset{+H^+}{\overset{-H^+}{\rightleftharpoons}} HPO_4^{2-}$$

$$H_3\overset{+}{N}.CH_2.COOH \underset{+H^+}{\overset{-H^+}{\rightleftharpoons}} H_3\overset{+}{N}.CH_2.COO^-$$

$$\underset{+H^+}{\overset{-H^+}{\rightleftharpoons}} H_2N.CH_2COO^-$$

[D:54]

8   In the case of a weak acid, the fraction $\alpha$ slowly increases with dilution and gradually approaches unity at high dilution. A strong acid has $\alpha = 1$ except at high concentrations.
[E:94]

9   Definitions:

$$pH = \log_{10}(1/[H^+]) = -\log_{10}[H^+]$$
$$pK_a = \log_{10}(1/K_a) = -\log_{10}K_a$$
$$pK_w = \log_{10}(1/K_w) = -\log_{10}K_w$$

[D:54]

10   Let $c$ = concentration (mol/l) of the dissolved substance. In each case $c = 10^{-3}$.

(a) HCl,       $pH = -\log c = 3$

(b) $H_2CO_3$,     $pH = \frac{1}{2}(pK_a - \log c) = \frac{1}{2}(6 + 3) = 4.5$

(c) NaOH, $pH = pK_w + \log c = 11$

(d) NaHCO$_3$, $pH = \frac{1}{2}(pK_a + pK_w + \log c)$
$$= \frac{1}{2}(6 + 11) = 8.5$$

Note how the calculation for the weak acid or base is related to that of the strong acid or base respectively.
[A:3]

# B Buffers

11 A buffer solution resists the change in pH that would otherwise occur when a small quantity of a strong acid or base (i.e. hydronium or hydroxyl ions) is added either to the solution in the absence of buffering components or to an equal volume of water.
[E:97]

12 A buffer solution must contain a proton acceptor and a proton donor, in addition to the aqueous solvent. As the buffer base and the buffer acid must *both* be present in reasonable concentration, this implies that a strong acid plus a strong base has no buffer action.
[E:98]

13 Dissolved components concerned are:
   (a) $CH_3.COOH$ and $CH_3.COO^-$ (e.g., as sodium acetate),
   (b) $H_2CO_3$ and $HCO_3^-$ (e.g., as sodium bicarbonate),
   (c) $H_2PO_4^-$ and $HPO_4^{2-}$ (e.g., both as their potassium salts).
   [A:5, 10]

14 Buffer action will be shown by (b), (c) and (d), the acid and base pairs being respectively
   (b) $HOCH_2CH_2\overset{+}{N}H_3$ and $HOCH_2CH_2NH_2$
   (c) $H_3\overset{+}{N}.CH_2.COOH$ and $H_3\overset{+}{N}.CH_2.COO^-$
   (d) $H_3\overset{+}{N}.CH_2.COO^-$ and $H_2N.CH_2.COO^-$
   [E:98]

15 The useful equation (the Henderson–Hasselbalch equation) is the equilibrium equation for the ionization reaction expressed in logarithmic form as follows:
$$pH = pK_a + \log_{10} \frac{[base]}{[acid]}$$
[E:97]

16 The 3 values are respectively:
   $pH = pK_a - 1$; $pH = pK_a$; $pH = pK_a + 1$
   [C:12]

17 Equal buffering towards added $H_3O^+$ and $OH^-$ occurs when the pH equals the $pK_a$ of the buffer acid.
   [E:97]

18 At the point where the pH equals the $pK_a$ of the buffer acid, the curve of pH against moles of added $H_3O^+$ (or added $OH^-$) is linear and it remains approximately linear over a pH range of almost 2 units (i.e., from $pK_a - 1$ to $pK_a + 1$). However, in the neighbourhood of both the latter pH values, the curvature increases markedly and the buffering effectiveness declines. However, some buffering action will still remain if the incipient pH change is in the direction towards the $pK_a$ value, even when the ratio [base]/[acid] is 20/1 (or 1/20).
   [D:152]

19 The respective buffer capacities are governed by the magnitude of
   (a) concentration of conjugate base;
   (b) concentration of conjugate acid.
   [D:153]

20 This is mainly because metabolic processes produce large quantities of $CO_2$ and this reacts with water forming carbonic acid. Smaller quantities of other acids, e.g., lactic, $\beta$-hydroxybutyric, sulphuric, etc. are also produced metabolically. Bases and non-metabolic acids may also enter the blood by adventitious or by natural processes.
   [D:150]

21 The potential increase in $[H^+]$ is partly taken up by suitable basic groups in haemoglobin. After transport to the lungs, the Hb is oxidized to $HbO_2$ which, as it is a stronger acid, releases protons. The latter interfere with the $HCO_3^-/H_2CO_3$ equilibrium, leading to the production of more $H_2CO_3$ and thus the expiration of $CO_2$. High $HCO_3^-$ concentration in plasma provides notable $HCO_3^-/H_2CO_3$ buffering action towards acids even at pH 7.4. The $HPO_4^{2-}/H_2PO_4^-$ buffer system also contributes to a small extent to the blood's buffering capacity.
   [A:7]

## C Acidosis and alkalosis

22 Assuming that plasma carbonic acid concentration is proportional to plasma carbon dioxide concentration, which in turn is proportional to the partial pressure of $CO_2$, then

$$pH = 6.1 + \log_{10} \frac{[HCO_3^-]}{a(P_{CO_2})},$$

where $a = 0.225$ if $P_{CO_2}$ is in kilopascals and $a = 0.030$ if $P_{CO_2}$ is in mmHg. If desired, $[HCO_3^-]$ may be replaced by $[\text{Total } CO_2] - a(P_{CO_2})$.
[D:155]

23 Acidosis and alkalosis are pathological states, of the whole body, in which the blood plasma possesses an abnormally high $[H^+]$ or an abnormally low $[H^+]$ respectively. The normal $[H^+]$ value is taken as 40 nmol/l (pH = 7.40) with a conventional range of normality from 45 nmol/l (pH = 7.35) to 35.5 nmol/l (pH = 7.45).
[A:1201]

24 The following tend to produce respiratory acidosis: constriction or obstruction of respiratory passages, emphysema, pneumonia, paralysis of respiratory muscles, depression of the respiratory centre by hypoxia or by drugs (e.g., morphine or barbiturates), and inspiration of air with a high $CO_2$ content.
[C:524]

25 Hyperventilation tends to produce respiratory alkalosis. It may be due to hysterical overbreathing, inadequately controlled mechanical respiration, stimulation of the respiratory centre arising from drugs (e.g., salicylates) and injury to the respiratory centre.
[C:524]

26 In blood, the reaction system

$$H_2O + CO_2 \rightleftharpoons H_2CO_3 \rightleftharpoons H^+ + HCO_3^-$$

is present and the action of erythrocyte carbonic anhydrase ensures that there is always rapid attainment of near equilibrium concentrations.

In respiratory acidosis $P_{CO_2}$ rises and, in consequence, $[H^+]$ and $[HCO_3^-]$ must also rise, although, because of its relatively high concentration, the increase in $[HCO_3^-]$ is proportionately small. The rise in $[H^+]$ is moderated by its reaction with haemoglobin:

$$H^+ + Hb \rightleftharpoons H(Hb)$$

Respiratory alkalosis produces the reverse effects.
[C:524]

27 Renal processes will have compensatory effects. Thus, respiratory acidosis will give rise to renal enhancement of plasma $[HCO_3^-]$ and to increased secretion of $[H^+]$ from tubular cells, resulting in elevated excretion rates of $H_2PO_4^-$ and $NH_4^+$.
[C:524]

28 Any interference with normal oxidation of fats or carbohydrates (or enhanced metabolism of sulphur and phosphorus compounds to give sulphuric and phosphoric acids, respectively) may produce acidosis. Diabetic ketoacidosis and hypoxia are 2 examples.

In addition, non-respiratory acidosis can arise from:

(a) $H^+$ retention due to renal disease or to ureterocolic anastomosis;

(b) loss of intestinal bases by aspiration, fistula or diarrhoea.

[C : 524]

29 Non-respiratory alkalosis may arise from the intake of large doses of bases (e.g., $NaHCO_3$), from vomiting (in particular with pyloric stenosis) and in other situations where loss of gastric juice may occur. Loss of plasma $K^+$ can lead to a cellular uptake of $H^+$ and thus to an extracellular alkalosis.
[C:524]

30 Consider the reaction scheme

$$H_2O + CO_2 \rightleftharpoons H_2CO_3 \rightleftharpoons H^+ + HCO_3^-$$

(a) Non-respiratory acidosis, caused by metabolic production of stronger acids (than $H_2CO_3$) or by renal defects, will result in a shift to the left with reduced $[HCO_3^-]$ and increased $P_{CO_2}$ accompanying the increased $[H^+]$. The raised $[H^+]$ will stimulate the respiratory centre producing respiratory compensation.

(b) Loss of stomach acid will have the resulting effect of shifting the reactions to the right with reduced $P_{CO_2}$ and increased $[HCO_3^-]$ accompanying the reduced $[H^+]$.
[I:95, 102, 105]

31  Considering the 3 processes:
(a) plasma buffer systems can adjust with great rapidity;
(b) respiratory compensation may follow fairly rapidly to offset a non-respiratory disturbance;
(c) renal compensation of a respiratory disturbance is slow to take effect.
Only renal compensation dispels $H^+$ (partly as $H_2PO_4^-$ and $NH_4^+$).
[C:523]

# AMINO ACIDS AND PEPTIDES

## A  Fundamental properties of amino acids

32  In general, an amino acid contains an acidic group such as $-COOH$ or $-SO_2OH$ and an amino group that might be a primary, secondary, or tertiary amino group. Thus (a), (b), (c), (f) and (g) are amino acids. Structure (d) represents an amide and, whilst (e) and (h) contain $-NH_2$ groups, they are not acids.
[C:14]

33  The amino acids *involved in protein structure* are all:
(a) carboxylic acids, i.e., contain the $-COOH$ (or $-COO^-$) group;
(b) $\alpha$-amino acids, i.e., the amino (or imino) group is attached to the $\alpha$-carbon atom;
(c) $\alpha$-amino acids with at least one H atom directly attached to the $\alpha$-carbon atom.
In addition, with the exception of proline, they all possess the $-NH_2$ (or $-\overset{+}{N}H_3$) group and, with the exception of glycine, they are all L-amino acids.
[A:16]

34  Glycine is very soluble in water. The solubility of the others depends on the character of the substituent R present at the $\alpha$-carbon atom. The

presence of groups like –OH, –COOH and –NH$_2$ tends to increase hydrophilic character, whilst hydrocarbon groups increase lipophilic character.
[B:5]

35  The structure is a dipolar or zwitter ion structure of type $H_3\overset{+}{N}.CHR.COO^-$.
[B:3]

36  As glycine is a symmetrical molecule, it has only one stereoisomer. The other 19 amino acids, all possessing an asymmetric carbon atom, exist in D and L forms. The forms that are incorporated into proteins all have the same stereochemical configuration, that of L-alanine. This does *not* imply that they are all laevorotatory in relation to plane-polarized light.
[B:15]

37  Fully protonated glycine $H_3\overset{+}{N}.CH_2.COOH$ is a stronger acid than acetic acid $CH_3.COOH$ because the positive charge of the $H_3\overset{+}{N}-$ group facilitates the ionization of the –COOH group.
[F:40]

38  The ionization stages are:

$H_3\overset{+}{N}.CH(CH_3).COOH \rightleftharpoons H_3\overset{+}{N}.CH(CH_3).COO^-$
$\qquad\qquad\qquad\qquad\qquad\qquad + H^+$
$H_3\overset{+}{N}.CH(CH_3).COO^- \rightleftharpoons H^+$
$\qquad\qquad\qquad\qquad + H_2N.CH(CH_3).COO^-$

The isoelectric point $pI = \frac{1}{2}(pK_1 + pK_2) = 6.02$
[E:101]

39  The ionization stages of aspartic acid are:

(a) $HOOC.CH_2.CH(\overset{+}{N}H_3).COOH \overset{-H^+}{\rightleftharpoons}$
$\qquad\qquad HOOC.CH_2.CH(\overset{+}{N}H_3).COO^-$

(b) $HOOC.CH_2.CH(\overset{+}{N}H_3).COO^- \overset{-H^+}{\rightleftharpoons}$
$\qquad\qquad {}^-OOC.CH_2.CH(\overset{+}{N}H_3).COO^-$

(c) ${}^-OOC.CH_2.CH(\overset{+}{N}H_3).COO^- \overset{-H^+}{\rightleftharpoons}$
$\qquad\qquad {}^-OOC.CH_2.CH(NH_2).COO^-$

The 3 $pK_a$ values correspond to stages (a), (b) and (c). There is, of course, only one $pI$ which equals $\frac{1}{2}(pK_1 + pK_2)$.
[E:102]

40 In the fully protonated form, lysine has the structure

$$\overset{+}{N}H_3 \qquad\qquad \overset{+}{N}H_3$$
$$| \qquad\qquad\qquad |$$
$$CH_2.CH_2.CH_2.CH_2.CH.COOH$$

Three H atoms ionize in the following order: the first from the carboxyl group, the second from the $-\overset{+}{N}H_3$ at the terminal carbon atom and the third from the $-\overset{+}{N}H_3$ at the $\alpha$-carbon atom.

The zwitter ion has structure

$$H_2N.CH_2.CH_2.CH_2.CH_2.CH(\overset{+}{N}H_3).COO^-.$$

[A:22]

41 Chemical synthesis of peptides from amino acids requires more than one reaction stage. In cells also, the amino acids need to be converted into more reactive derivatives before the delicate mechanisms for peptide synthesis operate.
[A:46]

42 The ninhydrin test produces a blue-purple colour with most $\alpha$-amino acids, $CO_2$ being produced in the reaction. Proline gives a yellow colour. A blue colour can also be produced by amines and by peptides.
[A:38]

B  **Properties of amino acid side chains** [*The amino acids included are the 20 commonly present in proteins*]

43 Each 3 letter amino acid symbol is used to represent either the whole $\alpha$-amino acid molecule or an amino acid residue present within, or at the ends of, a polypeptide chain. The required symbols are (a) Ile, (b) Asn, (c) Gln and (d) Trp.
[G:91]

44 In writing the primary structure for a long polypeptide chain, even the 3 letter symbols may be cumbersome and, for the discussion of such complex structures, a single letter abbreviation scheme has been developed. As the relationship of the single letters to the names of the amino acids is less obvious than for 3 letter abbreviations, their use is mainly confined to research papers.

Examples: Leucine L, Lysine K, Phenylalanine F, Tryptophan W.
[F:16]

45 The examples are:
   (a) aspartic acid and glutamic acid,
   (b) asparagine and glutamine,
   (c) arginine and lysine,
   (d) serine and threonine,
   (e) cysteine and methionine.
   [D:41]

46 (a) The molecule of glycine, having no asymmetric carbon atom, is symmetrical and does not exist in D and L forms.
   (b) Proline is strictly an imino acid having a secondary amino group NH. The presence of this group as part of a 5-membered ring means that, when proline is a residue in a polypeptide chain, the chain necessarily undergoes a sharp change in direction at that point.
   [A:23, 30]

47 The amino acids phenylalanine and tyrosine both possess a benzene ring. Tyrosine ($p$-hydroxyphenylalanine) has an –OH group substituted in the benzene ring. This is a phenolic –OH group. It must markedly reduce the lipophilic character that the phenyl group imparts to phenylalanine.
   [D:41]

48 Histidine is a weakly basic amino acid (isoelectric point $pI = 7.6$) because of the weakly basic character of the imidazole ring.
   [D:42]

49 The –SH groups, present in 2 cysteine side chains, are readily oxidized to give cystine, which possesses a disulphide bridge:

$$H_3\overset{+}{N}-CH-COO^-$$
$$|$$
$$CH_2$$
$$|$$
$$SH$$
$$+$$
$$\xrightarrow[\text{reduction}]{\text{oxidation}}$$
$$SH$$
$$|$$
$$CH_2$$
$$|$$
$$H_3\overset{+}{N}-CH-COO^-$$
2 cysteine molecules

$$H_3\overset{+}{N}-CH-COO^-$$
$$|$$
$$CH_2$$
$$|$$
$$S$$
$$|$$
$$S$$
$$|$$
$$CH_2$$
$$|$$
$$H_3\overset{+}{N}-CH-COO^-$$
1 cystine molecule

Disulphide bridges can link polypeptide chains.
[F:18]

50 The following are *selections* of 5 amino acids with
  (a) hydrophilic side chains: Ser, Thr, Asp, Glu, Asn,
  (b) lipophilic side chains: Val, Leu, Ile, Phe, Met.
  [C:16]

51 *Side chains* of the following cannot be involved in hydrogen bonding: Ala, Val, Leu, Ile, Met, Pro and Phe. In addition, the $CH_2$ group of glycine cannot participate in hydrogen bonding and the hydrogen-bonding propensity of the –SH group of cysteine is virtually zero.
  [B:6]

52 (a) The following have a second asymmetric carbon atom: Ile, Thr.
  (b) Trp and Tyr exhibit u.v. light absorption in the 270–280 nm region, but only tryptophan shows strong absorption. Trp is responsible for most of the u.v. light absorption of proteins.
  [C:16, 20]

## C  The peptide bond and polypeptides

53 Consider hypothetically 2 $\alpha$-amino acids combining together with the elimination of water between them:

$$H_3\overset{+}{N}.CHR^1.COO^- + H_2\overset{+}{N}H.CHR^2.COO^-$$

$$\Big\downarrow{-H_2O}$$

$$H_3\overset{+}{N}.CHR^1.CO-NH.CHR^2.COO^-$$

The C–N bond linking the CO and NH groups is an amide-type bond, which when linking two amino acid residues is called a peptide bond.
[F:17]

54 No, the peptide bond has partial double-bond character. As a consequence of loss of free rotation, the peptide group has a planar configuration as indicated:

[F:28]

55 Although glutathione is glutamyl–cysteinyl–glycine, it cannot be considered a typical tripeptide because the glutamyl–cysteine linkage takes place with the γ-carboxyl group of the glutamic acid residue.
[C:29]

56 A polypeptide has a direction or polarity because one end possesses an $H_3\overset{+}{N}-$ group and the other a $-COO^-$ group. Conventionally, they are always written with the amino group on the left and the carboxyl group on the right. Hence $(H_3\overset{+}{N})$ Ala–Gly–Ser–Ala $(COO^-)$ differs from $(H_3\overset{+}{N})$ Ala–Ser–Gly–Ala $(COO^-)$.
[F:17]

57 As polypeptides have a direction, the cyclic formula is incomplete without knowledge of whether the NH–CO direction is clockwise or anticlockwise. The correct direction is clockwise for the gramicidin S formula.
[F:663]

58 Covalent cross-links are formed by disulphide bridges linking cysteine residues.
[F:33]

59 The disulphide bond may be broken by treatment with oxidizing agents such as performic acid or hydrogen peroxide, which form cysteic acid residues.

$$-S-S- \rightarrow -SO_2OH + HOO_2S-$$

Reduction will also break the disulphide bond, but the –SH groups produced may readily reoxidize.
[F:32]

60 Complete hydrolysis to amino acids is effected by prolonged heating with 6M HCl in a sealed tube at 110 °C. All the tryptophan is destroyed in this hydrolysis, and several other amino acids undergo dehydration or cyclization. Tryptophan is obtained from hydrolysis with alkali. Enzymic hydrolysis may be relatively unspecific with some bacterial peptidases, or specific for certain peptide bonds with other proteolytic enzymes.
[C:26]

61 The structure of TRH is chemically related to the

tripeptide Glu–His–Pro. The glutamine residue has, however, cyclized with the elimination of water, and the carboxylic acid group of the proline has been converted into an amide group.
[C:29]

# INVESTIGATION TECHNIQUES

## A  Separation of subcellular fractions

62  The method must be appropriate for the type of tissue or cell and the purpose for which homogenization is required. The following methods are available: homogenization using blenders with rotating blades or with Dounce-type glass plunger homogenizers, extrusion under pressure, ultrasonic disintegration, alternate freezing and thawing, osmotic disruption, and breakdown with enzymes, solvents or other agents.
[G:318]

63  No. The centrifugal force acting on the particles should be calculated. At least the average radius of rotation (distance from the axis of rotation to the midpoint of the liquid column) should be quoted in addition to rotor speed and time.
[E:74]

64  The tissue homogenate is divided into fractions by centrifugation in several steps, the applied centrifugal force, and usually the time of centrifugation, increasing for each step. The centrifugal field for each stage is chosen to permit a particular cell organelle (or other material) to deposit as a pellet during the centrifugation time. The pellet is 'purified' by suspension in a 'washing' medium followed by centrifugation, whilst the supernatant proceeds to the next fractionation step.
[H:284]

65  Whole cells and cell debris sediment most readily, followed by mitochondria, lysosomes and peroxisomes. Ultracentrifugation is required to separate the microsomal (and ribosomal) fraction.
[G:318]

66  The sample is carefully placed in a thin layer onto the surface of a solution possessing a concentration, and hence a density, that increases with depth. The increase in density may be continuous or discontinuous. Centrifugation is then carried out.
[G:100]

67  The solute most commonly employed for density gradients is sucrose (buffered or unbuffered). Other materials sometimes used are caesium chloride, and rubidium chloride.
[C:38]

68  The technique of zonal centrifugation involves layering an appropriate sample over a solution possessing a continuous density gradient. Centrifugation for an appropriate time will effect separation of the components into discrete (but still moving) zones or bands.
[G:100]

69  The material is suspended in a liquid having a density equal to the mean buoyant density of the particles to be separated. After contrifugation for an appropriate time, lighter particles will have risen to near the meniscus, whilst heavier particles will be at the base of the tube. Isopycnic separations may also be effected using continuous density gradients.

70  The sample is dispersed in a concentrated CsCl solution. Centrifugation then produces a gradient because of the $Cs^+$ ion. The suspended particles will separate until they attain equilibrium positions at which the buoyant density of each fraction equals the density of the adjacent medium.

71  A marker enzyme is presumed to be located in just one subcellular site and, consequently, its activity in any fraction may be taken as an index of the proportion of a particular subcellular entity in the fraction. Cytochrome oxidase and monoamine oxidase are often used as mitochondrial marker enzymes. It is usual to assay for at least two marker enzymes in each fraction.

## B  Separation of molecular components

72  The main requirement for a preparative separation is that an adequate quantity of product(s)

should be obtained in an acceptably pure state. Recovery of the components does not have to approach 100%. In contrast, separations for assay purposes should be on a very small scale and be effected with no loss of components. However, once appropriate measurements have been made, it may not matter if the assayed components are not recovered.

73 Precipitation, dialysis, countercurrent extraction, and column chromatography are methods particularly suitable for preparative isolations, whilst gas-liquid, thin-layer, and affinity chromatography and electrophoresis on cellulose acetate are high resolution methods, particularly suitable for analysis (qualitative and quantitative).
[G:110]

74 Adsorption chromatography depends on the different adsorption characteristics of solutes onto a solid (finely powdered) adsorbant from (usually) a dry organic solvent. Many column and thin-layer methods using an organic solvent are of this type. In partition chromatography, separation depends on differences in the partition of solutes between two phases (usually aqueous and non-aqueous). Paper chromatography is of this type. Both adsorption and partition are involved simultaneously in many chromatographic systems.
[G:115]

75 An $R_F$ value is defined for practical purposes in chromatography as the distance moved by the separated component divided by the distance moved by the solvent front in the same time.
[H:273]

76 In g.l.c., each compound is distributed between a stationary liquid and a mobile gas phase. A liquid phase commonly employed is silicone grease, supported on inert grains in an inert column maintained at a suitable temperature. The vapour is carried through the column in an inert gas such as argon. As the gas phase leaves the column it enters a detector system.
[G:118]

77 Porous beads, usually made of 3-dimensional

polymers (cross-linked dextrans have the trade name Sephadex), form the stationary gel phase by uptake of aqueous solvent. The components in solution pass through a column of gel particles and whilst smaller molecules can readily enter the molecular matrix of each gel particle, larger molecules are excluded.
[G:122]

78   A suitable competitive inhibitor of the enzyme is modified chemically so that it can then react with an insoluble matrix and attach ligand groups that still retain their affinity for the active sites of enzyme molecules. The specificity of the enzyme interaction enables an affinity column to separate the enzyme readily from a protein mixture.
[G:124]

79   Ion exchange separations are usually carried out in columns packed with resin particles that contain either acidic groups (a cation exchanger) or basic groups (an anion exchanger). Ion exchange resin swells in a buffer solution and when ionic materials pass through a column of such a gel, separation occurs dependent on charge, molecular size and other properties related to the rate of diffusion of the ions through the resin matrix to the interior charged sites.
[G:119]

80   In both cases a membrane is used that permits water and small particles and ions to pass through freely, but retains macromolecules and colloidal aggregates. In dialysis, the membrane is placed between the colloidal fluid and water (buffered if necessary) so that small molecules are removed from the colloid by diffusion. In ultrafiltration, hydrostatic pressure, aided by applied pressure, extrudes water and dissolved molecules through the membrane (filter), concentrating the macromolecules in the retained colloid.
[F:18]

81   Most hydrophilic biological macromolecules possess ionizable groups and at a suitable pH are capable of migration in an externally applied electric field. As the rate of migration will depend on the magnitude of the charge and on

74

molecular size and shape, electrophoretic separations are possible. Ions of any size are capable of electrophoretic separation.
[D:58]

82 The following are examples of media used in electrophoresis: strips of paper or of cellulose acetate, thin layers of silica or alumina, and starch, agar and polyacrylamide gels.
[H:281]

## C  Colorimetry and spectrometry

83 No. The quantity called the extinction $E$ (or absorption) is proportional to the concentration of coloured component and could be used as a measure of concentration in relation to that glass cell. Extinction $E$ is defined as $\log_{10}(I_o/I)$.
[H:286]

84 According to the Beer–Lambert law, the extinction $E$ is proportional to the concentration $c$ of absorbing substance and to the thickness of the absorbing layer (or the path length) $l$, i.e., $E = \varepsilon\, c\, l$. Constant $\varepsilon$ is called an extinction coefficient for the absorbing solute.
[H:286]

85 Strong absorption of light by a solute takes place over a comparatively narrow wavelength range, and the light source of a colorimeter covers a wider range. As $\log_{10}(I_o/I)$ is the physical quantity provided by a colorimeter, greater accuracy as well as greater specificity is obtained by having $I_o/I$ as high as possible. This is partly achieved by limiting light intensity measurement to the spectral region in which strong absorption occurs.
[II:285]

86 Various operations, possibly involving reagents that are themselves coloured, are required to produce the coloured solution placed in the colorimeter. It is essential to carry out these operations not only on test samples, but also, in an identical way, on a sample from which the substance to be assayed is absent. The solution that results is called a 'blank'. It may be coloured. It must be used to set the zero of the instrument. Use of a blank helps to ensure that

the measured extinction (or absorption) arises only from the components being assayed.

87  The relations are $\lambda\nu = c$ and $\lambda\bar{\nu} = 1$, where $c$ is the velocity of light.
[E:205]

88  The relation is $\Delta E = h\nu$, where $h$ is Planck's constant. If the radiation of frequency $\nu$ is considered as a beam of photons, then $E = h\nu$ is the energy of a photon.
[E:205]

89  An elevation in the energy level of an outer electron (bonding or non-bonding) will give rise to absorption in the visible or ultraviolet region.
[E:206]

90  Near infrared absorption arises from increase in vibrational energy of a molecular structure. The vibrational energy may often be referred to a definite bond and may involve a stretching or a bending mode of vibration.
[E:206]

91  The emission consists of one or more lines from the atomic spectrum of an atom. The specific de-excitation of an excited outer orbit electron produces emission of a particular wavelength.

92  Other spectra of value for biochemists include fluorescence and Raman spectra, nuclear magnetic resonance and electron spin resonance spectra, and mass spectra. The latter differs from the others in not being an electromagnetic radiation spectrum.

## D  Radioisotopes

93  They are all stable (non-radioactive) isotopes.
[G:319]

94  Whereas $^{226}Ra$, in common with many other large radioactive nuclei, is an $\alpha$-particle ($^{4}He^{2+}$) emitter, $^{14}C$ and $^{32}P$ are $\beta$-particle (electron) emitters. Isotope $^{14}C$ is a weak $\beta$-emitter of comparatively long half-life (5730 years) whilst $^{32}P$ is a high energy $\beta$-emitter with a short half-life (14.3 days).
[F:295]

95 Isotopes $^3$H, $^{14}$C, $^{32}$P and $^{35}$S have proved very useful in research and $^{51}$Cr, $^{59}$Fe, $^{133}$Xe and $^{131}$I are examples of isotopes of use in clinical investigations.

96 Both γ-rays and X-rays are high energy electromagnetic radiation, but γ-rays have higher energy and lower wavelength than X-rays. γ-Rays originate from nuclear transformations, whereas X-rays originate from de-excitation of electrons to their ground state in the inner orbits of larger atoms.

97 This is a commonly used traditional (not SI) unit for the quantity of radioactive material (or activity) that gives the same number of disintegrations per second ($3.7 \times 10^4$) as 1.0 $\mu$g of radium.
[E:214]

98 As the decay of a radioactive isotope is exponential, the fraction that undergoes disintegration in any time interval is independent of the quantity initially present. Thus the time for half of any quantity of isotope to disintegrate is a constant for the isotope and is called its half-life.
[E:216]

99 Commonly used types are:
(a) Geiger–Müller counters, which depend basically on the ionization produced in a gas by the passage of α- and β-particles (the ionizing effect of γ-rays is very weak).
(b) scintillation counters, which depend on radiation (α, β, or γ depending on the scintillator) exciting certain substances resulting in scintillation, measurable by means of a photomultiplier.
In practice, scintillation counting often has many advantages over Geiger–Müller counting for the β-emitting isotopes used in biochemical research.

100 Some structures (e.g. hexose rings and benzene rings) may pass unchanged through a particular metabolic sequence and, in consequence, it would not matter which ring carbon atom is labelled with $^{14}$C or ring CH is labelled with $^3$H. In other circumstances, reaction would remove a label from one position and not from another.

Finally, in many structures, combined tritium (T–X) can often exchange readily with the light hydrogen isotope in suitable H–Y structures.
[F:295]

# PROTEINS, STRUCTURE AND FUNCTION

## A Sequencing of peptide chains

101 Any method of purifying macromolecules could be applied to proteins, but particularly important methods of general use in protein purification are: dialysis, fractional precipitation (by ammonium sulphate), gel filtration, ion exchange chromatography, and electrophoresis.
[G:110]

102 Polypeptide chains can be linked covalently or non-covalently. The main example of covalent linking is a disulphide bridge between cysteine residues. For sequencing purposes, a disulphide bridge is usually broken by oxidation with performic acid (or hydrogen peroxide). An important example of largely electrostatic bonding is hydrogen bonding, which forms the usual link between protein subunits. Hydrogen bonds between polypeptide chains may be broken by dispersing the protein in a concentrated (8M) urea solution.
[F:27, 30]

103 Amino acid mixtures are usually separated either by ion exchange chromatography on a sulphonated polystyrene column or by paper (or thin-layer) chromatography.
[F:22]

104 Fluorodinitrobenzene reacts with a terminal amino group producing a dinitrophenyl (DNP) derivative of the polypeptide, thus chemically labelling the N-terminal amino acid. Hydrolysis will give the constituent amino acids, but the N-terminal amino acid will still have the DNP group attached. After separation of the amino acids the DNP-amino acid can be identified.
[F:23]

105 An alternative reagent is dansyl chloride (5-dimethylaminonaphthalene-10-sulphonyl chloride). This reacts with terminal amino groups to form highly fluorescent sulphonamide-type derivatives.
[F:23]

106 Aminopeptidases specifically hydrolyse the first peptide bond at the N-terminal amino acid, whilst carboxypeptidases specifically hydrolyse the first peptide bond at the carboxyl terminal amino acid.
[C:27]

107 Trypsin specifically hydrolyses peptide bonds formed from the carboxyl group of lysine or arginine. Chymotrypsin is almost as specific and mainly hydrolyses peptide bonds formed from the carboxyl groups of tryptophan, tyrosine and phenylalanine.
[F:26]

108 Cyanogen bromide (BrCN) reacts with methionine residues, cleaving a polypeptide chain as follows:

[F:25]

109 In the Edman degradation, the peptide reacts with phenyl isothiocyanate at the N-terminal to give a derivative that, under mildly acidic conditions, splits off the terminal amino acid only to give a phenylthiohydantoin and a new peptide lacking the original N-terminal amino acid. This process can be repeated stepwise enabling each N-terminal amino acid to be identified in turn. The process has been made automatic in sequenator instruments.
[F:24]

110 After determining the number of amino acids of each type in the molecule, the Edman degradation could provide their sequence. However, the usual procedure is to hydrolyse the polypeptide

with various proteolytic enzymes and, if methionine is present, to divide the chain using cyanogen bromide. After sequencing the fragments, study of the overlap that arises from different cleavage positions usually enables the sequence for the whole polypeptide to be elucidated.
[A:42]

## B  Conformation of peptide chains

111 No enzyme is required. A polypeptide rapidly adopts the normal 3-dimensional configuration, which is that of minimum free energy in relation to the pH, ionic strength, temperature and other conditions of the medium (cytoplasmic or equivalent).
[F:34]

112 Non-covalent association may arise from hydrogen bonding, electrostatic attraction between oppositely charged ions, and hydrophobic interaction between aliphatic and/or aromatic substituents. The most common covalent type of association is the disulphide bond.
[C:32]

113 (a) The peptide bond possesses a partial double-bond character $C\!=\!\!=\!N$. This greatly inhibits free rotation about the bond and, in fact, makes the peptide group a rigid planar structure with the carboxyl oxygen usually *trans* to the NH hydrogen.

(b) Presence of the ring in proline makes the direction of a peptide chain deflect sharply through approximately a right angle at a proline residue.
[F:28, 14]

114 An $\alpha$-helix is formed by hydrogen bonding

$$\diagdown N\text{--}H\cdots O\!=\!C\diagup$$

within the chain. The NH group of residue $x$ is hydrogen bonded to the CO group of residue $(x-4)$ whilst the CO group of residue $x$ is hydrogen bonded to the NH group of residue $(x+4)$. Thus all NH and CO groups, within the tightly coiled structure, are appropriately joined to form a stable right-handed helical (rod-like)

structure with side chains extending radially
from the helix.
[F:29]

115 The β-structures are also produced by
N–H··O=C bonding, but between many
extended polypeptide chains all in one plane.
Alternate chains may be parallel (as shown) or
antiparallel.

$$
\begin{array}{c}
\vdots \\
\text{O} \\
\parallel \\
\text{–N–CHR–C–N–} \\
\mid \qquad\qquad \mid \\
\text{H} \qquad\qquad \text{H} \longrightarrow \\
\vdots \qquad\qquad \vdots \\
\text{O} \qquad\qquad \text{O} \longrightarrow \\
\parallel \qquad\qquad \parallel \\
\text{–C–N–CHR–C–} \\
\parallel \\
\text{H} \\
\vdots
\end{array}
\Biggr\} \; 2 \text{ parallel chains}
$$

[F:31]

116 The primary structure comprises the amino acid
sequences and the disulphide bridges of the polypeptides. The secondary structure is the spatial
arrangement of residues close together in a chain
and held together by hydrogen bonds (α-helix
and β-pleated sheet). Tertiary structure is the
overall arrangement of long segments of secondary structure and their connecting chains. Finally, 2 or more polypeptide chains may loosely
associate to give an aggregate of subunits. The
number and arrangement of these is called
quaternary structure.
[F:32]

117 Unfolding of the 3-dimensional tertiary and secondary structure without damage to the primary
structure (e.g., by 8M urea) will probably remove the protein's specific biological properties
(i.e. denaturation). However, removing the unfolding reagent, and restoring the protein to its
normal aqueous milieu can produce spontaneous
refolding and re-establishment of activity.
[F:34]

118 The main approach for establishing 3-
dimensional structure is X-ray crystallography.
Many other physical methods may have a sub-
sidiary value, e.g. spectroscopic methods, including nuclear magnetic resonance; optical

rotatory dispersion and circular dichroism; and electron microscopy for the overall shape of large proteins.
[F:45]

## C Types of protein

119 Albumins are more soluble than globulins. They disperse readily in water and require saturation with ammonium sulphate for precipitation. Globulins do not disperse in distilled water, but disperse in dilute salt solutions. They are precipitated by ammonium sulphate at half saturation.
[C:31]

120 Scleroproteins are proteins, such as keratin, elastin and collagen, that have low solubility. They fail to disperse in water, in dilute salt solutions and in alcohol.
[C:31]

121 Histones are basic proteins found in association with DNA. The separated histones are soluble in water and in dilute acids and alkalis. They will not disperse in dilute aqueous ammonia.
[C:352]

122 Protein is associated with carbohydrate in glycoproteins, with lipid in lipoproteins and with nucleic acid in nucleoproteins.
[D:52]

123 Ferritin is a protein containing combined $Fe^{3+}$, alcohol dehydrogenase is a protein that contains $Zn^{2+}$ and ceruloplasmin is a protein containing combined $Cu^{2+}$.
[C:83]

124 Haemoglobin, chlorophyll and rhodopsin are examples of highly coloured proteins.
[D:52]

125 (a) Phosphoproteins are phosphate esters of serine or threonine residues present in the polypeptide chains. Some enzymes are phosphoproteins.
(b) Flavoproteins contain attached groups (prosthetic groups) which may be flavin mononucleotide (FMN) or flavin–adenine dinucleotide (FAD). Flavoproteins are involved

in certain dehydrogenation processes.
[C:300, 100]

126 (a) In insulins, almost half of the amino acid
residues in the molecule can be replaced by
other amino acids and the molecule will retain its insulin activity. The nature and the
position of the remaining 26 amino acids are
essential for activity.

(b) In immunoglobulin (IgG) molecules an extensive region of the molecules is invariant.
However, if portions of the chains at the
N-terminal ends are altered, immunological
properties will be preserved, but the particular antigen specificity will probably change.
[C:475, 545]

# OXYGEN-TRANSPORTING PROTEINS, HAEM AND PORPHYRINS

## A  Haem and haemoproteins

127 The haem group contains an $Fe^{2+}$ linked at the
centre of a planar heterocyclic ring structure
called protoporphyrin. This consists of 4 pyrrole
groups joined by 4 CH bridges with the 4 cyclic
N atoms directed towards the centre. Side
groups are attached to the outer cyclic C atoms
and the whole ring structure contains extensive
double-bond conjugation.
[D:68]

128 Four bonds to the pyrrole N atoms lie approximately in the plane of the porphyrin ring and 2
further linkages are possible that act in opposite
directions to each other and at right angles to the
plane of the ring. Groups attached by the latter 2
linkages are said to be at positions 5 and 6.
[D:68]

129 In a haemoprotein, a haem group is attached to a
polypeptide chain by either weak or strong covalent bonds. The haemoprotein may be monomeric
or contain haemoprotein subunits linked together.
[C:306]

130 The following are examples of haemoproteins:
myoglobin, haemoglobin, cytochromes, perox-
idase, catalase and tryptophan pyrrolase.
[A:767]

131　The association is weak in haemoglobin, being largely due to an $(Fe^{2+})$–N linkage at position 5 with a proximal histidine residue (F8). A distal histidine residue (E7) is present, but at a greater distance from the $Fe^{2+}$. In cytochromes, not only are there linkages with amino acids at both positions 5 and 6, but the ring side chains may be covalently bound to the protein (through cysteine residues in cytochrome c).
[A:243]

132　Treatment of myoglobin with dilute HCl will produce dissociation.
[F:56]

133　In haemoglobin, the position 6 accepts an oxygen molecule (between the iron and the distal histidine residue) to form oxyhaemoglobin. The iron remains throughout in the $Fe^{2+}$ state. Thus the oxidation–reduction is:

$$Hb + O_2 \rightleftharpoons HbO_2$$

In the cytochromes, oxidation occurs by $Fe^{2+}$ losing an electron. The essential oxidation–reduction process is:

$$Fe^{2+} \rightleftharpoons Fe^{3+} + e^-$$

[A:244]

134　Haemoglobin may be oxidized by ozone, nitrites and other oxidizing agents to methaemoglobin, where the iron is in the $Fe^{3+}$ state. Methaemoglobin cannot function as an oxygen carrier.
[A:762]

135　In normal erythropoiesis, haemoglobin is synthesized in the nucleated erythrocytes of the bone marrow and in reticulocytes. It is not synthesized or degraded during the 120–135 day life span of the circulating mature erythrocyte.
[A:777, 782]

136　In a 70 kg man, the normal daily turnover of haemoglobin is about 6 g.
[A:317]

137　Haem is synthesized in most tissues. Apart from its synthesis in bone marrow for the production of haemoglobin, it is required by other tissues for

the formation of cytochromes and other haemo-proteins. The liver is the largest non-erythro-poietic producer of haem. As the initial and final parts of the synthesis of haem from glycine and succinyl CoA take place in mitochondria, only cells containing mitochondria can synthesize haem.
[C:307]


**B  Porphyrins and porphyria** [In questions and answers 138 to 151, the word porphyrin is abbreviated to (pn)]

138   Proto(pn) III (or IX) has structure:

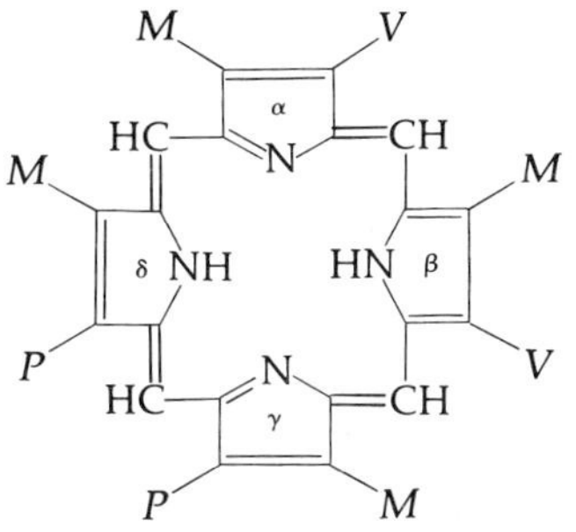

where $M = -CH_3$
$V = -CH=CH_2$
$P = -CH_2.CH_2.COO^-$


It contains 4 pyrrole rings $\alpha$, $\beta$, $\gamma$, and $\delta$, linked by 4 CH bridges in a fully conjugated structure, with the N atoms in the central region of the molecule and with substituent methyl, vinyl and propionate groups round the periphery. It is the precursor of haem, which forms when the NH hydrogens on rings $\beta$ and $\delta$ are replaced by a central $Fe^{2+}$. Coordinate bonds link this $Fe^{2+}$ to the N atoms in rings $\alpha$ and $\gamma$.
[A:771]

139   Succinyl CoA reacts with glycine in the presence of pyridoxal phosphate and $Mg^{2+}$. Coenzyme A and then $CO_2$ are eliminated and $\delta$-amino-laevulinic acid (ALA) is formed. The enzyme is ALA synthetase and production of ALA is the rate-controlling step in porphyrin synthesis.

$$HOOC.CH_2.CH_2.COS(CoA) + H.CH(\overset{+}{N}H_3).COO^-$$

$$\downarrow$$

$$HOOC.CH_2.CH_2.CO.CH_2NH_2$$

ALA dehydrase catalyses the condensation of 2 ALA molecules, with elimination of $2H_2O$, to produce porphobilinogen (PBG).

where $A = -CH_2.COO^-$
$P = -CH_2.CH_2.COO^-$

(PBG)

ALA is formed in mitochondria, but passes into the cytosol, where PBG is synthesized.
[A:768, 773]

140  Four PBG molecules combine in the cytosol. Four $NH_4^+$ ions are eliminated and uro(pn)ogen III is produced with a small proportion of uro(pn)ogen I. The uro(pn)ogens have structures:

$A = -CH_2.COO^-$
$P = -CH_2.CH_2.COO^-$
$X = A$ and $Y = P$ in uro(pn)ogen I
$X = P$ and $Y = A$ in uro(pn)ogen III

Uro(pn)ogen III is a precursor of haem whilst uro(pn)ogen I has no known function.
[A:770]

141 The haem synthetic sequence continues:

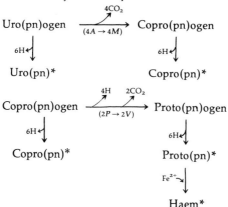

The reactions that produce proto(pn)ogen are purely transformations in the side chains, $4A \rightarrow 4M$ and then $2P \rightarrow 2V$. The 3 vertical reactions producing the (pn)s are oxidations increasing conjugation in the ring system. The final 3 stages in haem synthesis occur in mitochondria.
[A:771]

142 Highly coloured (dark red) compounds are given asterisks in answer 141. These compounds fluoresce in ultraviolet light.
[C:312]

143 ALA, PBG, uro(pn) and uro(pn)ogen all contain at least 3 hydrophilic groups. All these compounds are soluble in water and are excreted in urine. Proto(pn) and Proto(pn)ogen are much less soluble and excreted, *via* the bile, in faeces. Copro(pn) and copro(pn)ogen adopt an intermediate position, with urinary excretion increasing if pH is high. Screening tests on normal urine fail to demonstrate the presence of these compounds in urine. Traces of porphyrins are often readily demonstrable in normal faeces.
[C:313]

144 Porphyrias arise from a deficiency, usually inborn, of an enzyme of the haem synthetic pathway. This results in reduced feedback inhibition of the key ALA synthetase so that haem production may not be significantly reduced, but there is overproduction of some precursors.
[C:313]

145 Neurological symptoms, which may be abdominal pain and/or peripheral neuritis, are associated with high levels of plasma ALA and PBG.
[I:397]

146 Raised circulating levels of (pn)ogens and (pn)s produce skin sensitivity that varies from mild photosensitivity to severe blistering. This is due to high absorption of visible and ultraviolet light by (pn) structures.
[I:397]

147 Acute phases are precipitated by several types of drugs, e.g., oestrogens, barbiturates and sulphonamides, and also by other acute disease.
[I:400]

148 The three conditions are:
(a) acute intermittent porphyria,
(b) hereditary coproporphyria, and
(c) porphyria variegata.
They are all dominant inborn errors, all exhibit acute and latent phases and all show abdominal and/or neurological symptoms in the acute phase only. Skin lesions are absent in (a), unusual and mild in (b), and marked in (c). Enzymes that are deficient are those concerned with the uro(pn)ogen formation for condition (a), the proto(pn)ogen formation for condition (b), and haem formation for condition (c).
[I:399]

149 These are 2 rare inborn errors associated with the accumulation of porphyrins in erythrocytes. Acute attacks do not occur, but photosensitivity is present in both conditions.
[I:401]

150 Cutaneous hepatic porphyria is an acquired condition that is associated with marked photosensitivity. It occurs with liver disease consequent upon alcohol abuse and iron overload (Bantu siderosis).
[I:401]

151 Increased urinary or faecal excretion of (pn)s may be a secondary consequence of other conditions such as lead poisoning or liver disease.
[I:401]

# C  Myoglobin and haemoglobin

152 The myoglobin molecule contains a haem group
attached to protein. The protein consists of a
single polypeptide chain, of molecular weight
about 17 500, folded to form a very compact
globular structure. Eight right-handed $\alpha$-helices
make up about 75% of this structure and proline
residues, by changing the direction of the chain,
contribute to its compactness. Two His residues
are the only polar residues in the interior.
[F:62]

153 The haem group is located in a crevice between 2
histidine residues, the iron being bound at its
position 5 to the His residue F8. The propionate
groups of the haem reach the surface of the
whole molecule, but otherwise the haem is surrounded by non-polar groups. In oxymyoglobin
$MbO_2$, an oxygen molecule combines with the
iron atom at position 6.
[F:50]

154 The nearly spherical HbA structure is made up
of 4 polypeptide chains, each containing, within
a crevice, a haem group with its oxygen-
combining site. The 4 chains, consisting of 2
$\alpha$-chains (identical) and 2 $\beta$-chains (identical),
are held together by weak (non-covalent) forces
to form $\alpha_2\beta_2$. Although the $\alpha$- and $\beta$-chains differ
considerably from the myoglobin chain in amino
acid sequences, they are both similar to it confor-
mationally in the presence of a non-polar interior
and a crevice with 2 His residues.
[F:57]

155 Haemoglobin
   (a) reacts reversibly with, and transports, $O_2$,
       $CO_2$, and $H^+$,
   (b) is an allosteric protein in that uptake of an
       $O_2$ molecule at a haem group is regulated in
       an allosteric manner by the concentrations of
       $O_2$, $CO_2$, $H^+$ and of some organic phos-
       phates, these molecules acting at different
       sites.
   [F:82]

156 The sigmoid curve indicates cooperativity in the
binding of oxygen to haemoglobin. The uptake of
one molecule of oxygen at one site produces

slight distortion of the tetrameric structure that makes subsequent addition of the second, third, and fourth oxygen molecules progressively easier.
[F:65]

157 Increases in $[H^+]$ and $[CO_2]$ produce increased uptake of $H^+$ and $CO_2$ respectively by haemoglobin. Allosteric effects are then transmitted to oxygen binding sites and the oxygen affinity is reduced. The reverse occurs with decrease in $[H^+]$ and $[CO_2]$. The overall result is

$$HbO_2 + H^+ + CO_2 \underset{\text{in alveoli}}{\overset{\text{in tissues}}{\rightleftharpoons}} H^+Hb\text{–}CO_2 + O_2$$

[F:69]

158 One DPG anion fits in the central cavity of the Hb tetramer and is held by surrounding positively charged amino acid residues. It is released on oxygenation. Its presence reduces the oxygen affinity of the haemoglobin by a factor of 26 and this is essential for the release of adequate quantities of oxygen in tissue capillaries.
[F:70]

159 The DPG content of erythrocytes rises when arterial $P_{O_2}$ is abnormally low, for example in obstructive pulmonary emphysema or at high altitude. This results in an improvement of oxygen delivery to tissues under such abnormal conditions.
[F:71]

160 The tetrameric structure of haemoglobin imparts allosteric properties that the monomeric myoglobin lacks. Thus the binding of oxygen to haemoglobin is cooperative. The affinity for oxygen also depends on pH and on $P_{CO_2}$ and it is further regulated by DPG.
[F:65]

161 Whereas HbA contains 4 $\alpha$- and 4 $\beta$-chains, HbF contains 4 $\alpha$- and 4 $\gamma$-chains. The $\beta$- and $\gamma$- chains have the same number of amino acids, but have some sequence differences. The binding of DPG to HbF is weaker than to HbA and, consequently, oxygen has a greater affinity for HbF than for HbA. This ensures the appropriate

transfer of oxygen from maternal to fetal circulation.
[F:72]

# D Haemoglobinopathies

162 Various (early) embryonic haemoglobins contain
$\alpha$-, $\gamma$-, $\varepsilon$- and $\zeta$- chains, the synthesis of $\alpha$- and
$\gamma$-chains soon accelerates, whilst that of $\varepsilon$- and
$\zeta$-chains ceases. HbF ($\alpha_2\gamma_2$) then becomes the
main haemoglobin during the prenatal period.
After birth, synthesis of $\alpha_2\gamma_2$ normally declines
and HbF is replaced by HbA ($\alpha_2\beta_2$). A minor
adult type HbA$_2$ ($\alpha_2\delta_2$) together with traces of
HbF make up about 5% of the total haemoglobin
in normal adults.
[A:725]

163 $\alpha$-Chains contain 141 amino acids, whilst $\beta$-, $\gamma$-
and $\delta$-chains each contain 146 amino acids.
[A:728]

164 Apart from a hereditary persistence of fetal
haemoglobin, HbF level is raised in most $\beta$-tha-
lassaemias and in some other abnormal conditions.
[A:725]

165 Appropriate examples are:
   (a) haemoglobin S (HbS), whose presence in the
       homozygote causes sickle cell anaemia,
   (b) methaemoglobin (HbM) in which substitu-
       tion of the proximal or distal histidine by
       tyrosine stabilizes the haem group in the
       non-oxygen-binding ferric form. Hetero-
       zygotes are usually cyanotic.
   [F:92, 97]

166 Haemoglobin abnormalities may arise by amino
acid substitutions that distort the tertiary structure, or that interfere with the quaternary structure, so that allosteric interactions are modified.
Another type of abnormality may arise from an
imbalance in the synthesis of $\alpha$- and $\beta$-chains
(or of the other normal chains). This results in
the formation of abnormal quaternary forms such
as $\beta_4$ in haemoglobin H disease.
[F:97, 709]

167 The usual chain difference lies in a single amino

acid substitution. Over 100 variants are known and many are benign or give only mild symptoms, even in the homozygote.
[A:743]

168 Substitution of Glu 6 by a valine residue in each $\beta$-chain of HbA produces an HbS molecule.
[F:92]

169 Presence of non-polar Val residues on the outside of HbS markedly reduces solubility, but has much less effect on the oxygenated form. The deoxy form tends to separate in long fibrous precipitates that can distort the cell. Precipitation is favoured in circumstances where oxygen concentration is low.
[F:92]

170 In Africa, individuals with sickle cell trait appear to possess a protective mechanism against malaria infection.
[F:95]

171 Thalassaemias are a group of anaemias that arise from the reduced rate of synthesis of one or more haemoglobin peptide chains. They are produced by mutations affecting regulator genes.
[F:710]

# ENZYMES

## A Classification and general properties

172 (a) *Oxidoreductases* catalyse oxidoreductions between 2 substrates: e.g., L-lactate: NAD oxidoreductase (lactate dehydrogenase or LDH) catalysing the reaction

$$CH_3.CHOH.COO^- + NAD^+$$
$$= CH_3.CO.COO^- + NADH + H^+$$

(b) *Transferases* catalyse the transfer of a group from one substrate to another: e.g., ATP: D-hexose 6-phosphotransferase (hexokinase) catalysing

$$ATP + hexose = hexose\ 6\text{-}phosphate + ADP$$

(c) Hydrolases catalyse hydrolytic reactions: e.g., acetylcholine acyl-hydrolase (acetylcholinesterase) catalysing the reaction

$$CH_3.COOCH_2.CH_2.\overset{+}{N}(CH_3)_3 + H_2O$$
$$= CH_3.COOH + HOCH_2.CH_2.\overset{+}{N}(CH_3)_3$$

[D:98]

173 The 3 remaining enzyme classes are as follows:
(d) *Lyases* are non-hydrolytic enzymes that catalyse the splitting of a group from a substrate with the consequent formation of a double bond: e.g., L-malate hydro-lyase (fumarase) catalysing the reaction

$$HOOC.CHOH.CH_2.COOH$$
$$= HOOC.CH=CH.COOH + H_2O$$

(e) *Isomerases* catalyse isomerizations of any type: e.g., D-glyceraldehyde 3-phosphate ketol-isomerase (triosephosphate isomerase) catalysing the reaction

$$OHC.CHOH.CH_2OPO_3^{2-}$$
$$= HOCH_2.CO.CH_2OPO_3^{2-}$$

(f) *Ligases* catalyse the formation of linking bonds between two groups, with the concomitant breaking of a pyrophosphate linkage in ATP or some similar molecule: e.g., L-glutamate: ammonia ligase (glutamine synthetase) catalysing the formation of the CO–NH bond in the reaction

$$Glu + NH_3 + ATP = Gln + ADP + P_i$$

[D:98]

174 Urease was the first enzyme to be crystallized (by Sumner in 1926) and it was shown to be a protein. Since then, hundreds of enzymes have been crystallized. Note, however, that crystallization does not automatically imply that homogeneous (pure) enzyme protein has been obtained.
[G:196]

175 Although some enzymes (e.g., ribonuclease, pepsin, trypsin and lysozyme) are composed solely of polypeptide chains, many others require the association of some other ligand (prosthetic

group with the non-active protein (apoenzyme) to produce the active enzyme. Many prosthetic groups consist of, or contain, metal ions. Examples are $Mg^{2+}$ in most phosphotransferases, $Cu^{2+}$ in some oxidoreductases and $Zn^{2+}$ in carboxypeptidase A. Specific heterocyclic molecules must be associated with the enzyme protein in other cases. Such a non-protein structure (often called a coenzyme) may act as one of the substrates in the enzyme-catalysed reaction.
[G:202]

176 A transient state exists when the concentrations of initial reactants, intermediates, and products vary with time. This would occur in a time interval immediately following a mixing together of the reactants. A steady state exists where initial reactants are being converted into products, but the concentrations of intermediates remain constant with time. A transient state may tend to become a steady state as time progresses. Equilibrium exists when the concentrations of all reacting components are constant with time and when the velocities of forward and backward reactions are equal, i.e. there is no net reaction. A transient state may tend to become an equilibrium state with the passage of time.
[E:16]

177 (a) Consider a general enzyme-catalysed reaction sequence

$$\alpha A + \beta B + E \rightleftharpoons \text{intermediates} \rightleftharpoons$$
$$\gamma C + \delta D + E.$$

If the reaction velocity $v$ is to be defined in terms of reactant A, then

$$v = -\frac{d[A]}{dt} = -\left(\frac{\alpha}{\beta}\right)\frac{d[B]}{dt}.$$

If the sequence is at the steady state, then

$$v = -\frac{d[A]}{dt} = -\left(\frac{\alpha}{\beta}\right)\frac{d[B]}{dt}$$
$$= \left(\frac{\alpha}{\gamma}\right)\frac{d[C]}{dt} = \left(\frac{\alpha}{\delta}\right)\frac{d[D]}{dt}.$$

The units employed might be micromoles per litre per minute. Measurement of $d[D]/dt$ is often preferable to measurement of $-d[A]/dt$.

94

(b) The enzyme activity of a preparation is defined in terms of units of enzyme activity $U$. One unit of enzyme activity is defined as that quantity of enzyme preparation that transforms 1.0 micromole of substrate per minute in the presence of an excess (and optimum) substrate concentration, at optimum pH, and under other optimum conditions, but at a temperature of 30 °C (unless the latter temperature is specificially inappropriate.

(c) The specific activity of an enzyme is given as enzyme units per mg of protein.

[D:112, 136]

178 If the *actual mechanism* of a *single reaction* is accurately portrayed by the relation

$\alpha A + \beta B \rightarrow$ products,

then the reaction velocity will obey the equation

$v = k[A]^\alpha[B]^\beta$

where $\alpha$ and/or $\beta$ may have values 0, 1, 2, 3. The number $\alpha + \beta$ is called the order of the reaction and normally has values 0 (zero order reaction), 1 (first order reaction) or 2 (second order reaction). As the actual mechanism of any particular reaction can only be discovered by experimental investigation, the order of a reaction is determined from reaction velocity measurements at a range of substrate molarities. If the reaction has more than one stage, then a more complex function of concentration(s) may be required to give the velocity $v$.

[E:133]

179 Any single reacting act is only possible when the appropriate molecules collide in such a way that reacting groups are brought together. Even then, only a small proportion of collisions result in reaction because, for reaction to occur, the energy available at the instant of collision must be greater than a certain minimum. On a molar scale the magnitude of this energy barrier is called the energy of activation $E$ for the reaction. The greater the energy of activation, the higher is the energy barrier and the slower is the reaction.

A very slow reaction, however, may be dramatically accelerated if the reaction mechanism is changed. A single stage with high $E$ may be replaced by 2 or more stages each with a low

*E* value. The overall velocity is now governed by the *E* value for the least rapid stage in the new mechanism. Introduction of an appropriate enzyme alters the reaction mechanism with the results described.
[G:199]

180   Any enzyme-catalysed reaction carried out *under particular conditions* will have an optimum temperature at which the enzyme exhibits a maximum velocity. At temperatures well below this, the reaction velocity increases rapidly with temperature increase (approximately doubling for a 10 °C temperature increase). As the temperature nears the optimum, denaturation of the enzyme begins and at higher temperatures, denaturation is extensive, with rapid loss in activity. The extent of denaturation depends also on the time of incubation at 'elevated' temperatures.
[C:77]

181   At extreme pH values, enzyme denaturation will produce loss of activity. In narrower ranges of pH, the protonated state of active site groups of the enzyme or of substrate groups may influence the activity. This commonly leads to a bell-shaped curve, although the position of the activity maximum may vary widely, e.g. at pH 2 for pepsin whilst at pH 8 for trypsin (both proteolytic enzymes). Sigmoidal curves occur when the activity is insensitive to pH over several pH units, but then starts to decline rapidly when a particular pH is reached.
[C:77]

182   There are many examples of enzymic introduction of an asymmetric centre. It can occur because of asymmetry in the arrangement of 3 different functional groups at the active site of the enzyme. These can only be approached by the substrate molecule from one direction and so the substrate can only attach to the active site in a single way in spite of the symmetry of the substrate molecule. It follows that, although both asymmetric enantiomers appear equivalent from consideration of the substrate alone, only one will be synthesized by such an enzyme.
[C:53]

# B Mechanism of action

183 Enzyme E reacts reversibly with substrate A to give a complex EA, which can then break up reversibly to give E and product(s), the latter represented below by a single product P. If the reverse reaction involving P is not inherently very slow, then study of the overall process must be carried out under conditions when [P] is very small (initial velocity measurements) so that effectively

$$E + A \rightleftharpoons EA \rightarrow E + P$$

[F:110]

184 If $V$ is the maximum velocity approached at high [A] values and $K_m$ is the Michaelis constant for the enzyme-catalysed reaction, then

$$v = \frac{V[A]}{K_m + [A]}$$

or

$$\frac{V}{v} = 1 + \frac{K_m}{[A]}$$

[F:112]

185 The respective orders of reaction are (a) approximately first order; (b) approximately zero order.
[F:112]

186 The $v$ against [A] plot is a hyperbola.
[F:111]

187 $[A] = K_m$ when [A] is the substrate concentration that gives a reaction velocity $v$ equal to half the maximum velocity $V$.
[F:112]

188 A Lineweaver–Burk plot is a plot of $1/v$ against $1/[A]$, i.e., (velocity)$^{-1}$ plotted against (substrate concentration)$^{-1}$. A straight line plot shows that Michaelis–Menten kinetics are obeyed by the reaction.
[F:113]

189 In the 'ping-pong' mechanism, the enzyme exists in 2 forms $E_1$ and $E_2$, the active sites differing in the 2 forms. Considering the substrates A and B,

only substrate A reacts with $E_1$ and only substrate B reacts with $E_2$. With products P and Q, and in its simplest form, the complete mechanism becomes:

As with single substrate reactions, initial velocity measurements will make product-producing reactions effectively irreversible.
[D:115]

190   The reaction mechanism is as follows:

$$LDH \underset{}{\overset{+NAD^+}{\rightleftharpoons}} LDH\text{–}NAD^+ \overset{+lactate}{\rightleftharpoons}$$
(enzyme)                    (complex 1)

$$\qquad\qquad LDH\text{–}NAD^+\text{–}lactate$$
$$\qquad\qquad\text{(complex 2)}$$

$$\rightleftharpoons LDH\text{–}NADH\text{–}H^+\text{–}pyruvate \overset{-pyruvate}{\longrightarrow}$$
(rearranged complex 2)

$$\qquad\qquad LDH\text{–}NADH\text{–}H^+$$
$$\qquad\qquad\text{(complex 3)}$$

$$\overset{-H^+-NADH}{\longrightarrow} LDH$$
(enzyme)

It is an ordered mechanism, because the $NAD^+$ molecule attaches to the enzyme first followed by the lactate, in that order. The products are also relinquished in order.
[D:115]

191   A set of initial velocities at different substrate concentrations must be determined. Lineweaver–Burk plots of $v^{-1}$ against $[A]^{-1}$ for each value of [B] will give a family of parallel straight lines if the mechanism is ping-pong and non-parallel straight lines if the mechanism is ordered.
[E:187]

192   The active site of chymotrypsin contains a serine residue (number 195), a histidine residue (57), and a 'non-polar' region. The latter helps to provide the observed specificity of the enzyme, whilst residues 195 and 57 together act to cleave the specific –CO–NH– bond, liberating one of the

98

peptide products and forming a (peptide) acyl derivative at the serine (195) residue. Water then splits this acyl derivative to liberate the other peptide fragment.
[F:160]

# C 'Classical' inhibition

193 A reversible inhibitor is completely removable from the enzyme by dialysis and, if other conditions are unaltered, the enzyme activity will be unimpaired. An irreversible inhibitor damages the active site (e.g., it may become tightly bound to the enzyme by one or more covalent bonds) and is not removable by dialysis to leave the enzyme with its activity restored.
[F:116]

194 Both competitive inhibitor and substrate molecules can form complexes reversibly at the active sites, but each active site can only accommodate one of these molecules. In consequence, substrate and inhibitor compete with each other for the available active sites and hence the name competitive inhibitor.
[F:117]

195 A non-competitive inhibitor molecule combines reversibly with an enzyme molecule in such a way that no interference results with the attachment of a substrate molecule (A) to the enzyme's active site. This would normally imply attachment of inhibitor molecule to another site. The inhibiting action arises because the presence of the inhibitor (I) prevents direct breakdown of the complex (EAI) to give the products. The inhibitor decreases the turnover number of the enzyme.
[F:119]

196 An uncompetitive inhibitor molecule does not attach itself to the particular variant of enzyme molecule with which the substrate molecule reacts. In the case of a single substrate reaction, the uncompetitive inhibitor molecule forms a complex with the already formed enzyme-substrate complex and not with the enzyme itself. The complex EAI cannot dissociate to yield the reaction product(s) P.
[A:110]

197 If I is a competitive inhibitor:

$$E \overset{+A}{\rightleftharpoons} EA \longrightarrow E+P$$
$$+I \big\Updownarrow$$
$$EI$$

If I is a non-competitive inhibitor:

$$E \overset{+A}{\rightleftharpoons} EA \longrightarrow E+P$$
$$+I \big\Updownarrow \qquad \big\Updownarrow +I$$
$$EI \overset{+A}{\rightleftharpoons} EAI$$

If I is an uncompetitive inhibitor:

$$E \overset{+A}{\rightleftharpoons} EA \longrightarrow E+P$$
$$+I \big\Updownarrow$$
$$EAI$$

[A:109]

198 Yes. The apparent effect is that of a reaction being catalysed by a less active enzyme, but still operating by a Michaelis–Menten mechanism. Hence the term classical inhibition used to head this section.
[A:112]

199 All the Lineweaver–Burk (double reciprocal) plots will be straight lines. The cases are as follows:
   (a) *competitive inhibition*: the 2 lines will cross on the $v^{-1}$ axis. It follows that the maximum velocity $V$ is unaltered by the presence of the inhibitor.
   (b) *non-competitive inhibition*: the 2 lines will cross on the $[A]^{-1}$ axis and hence the $K_m$ is unaltered by the presence of the inhibitor. This picture is also given by a rapidly acting, irreversible inhibitor.
   (c) *uncompetitive inhibition*: the 2 lines will be parallel.
   [A:111]

200 Consider a competitive or a non-competitive inhibitor acting on a relevant enzyme-catalysed reaction in the steady state, then the inhibitor constant $K_I$ is defined as $[E][I]/[EI]$. No $E+I$ reaction occurs with an uncompetitive inhibitor,

but it is possible to define an analogous constant $K'_I$ as $[EA][I]/[EAI]$. The appropriate constants $K_I$ and $K'_I$ may be calculated from measurements taken from Lineweaver–Burk plots for the inhibited and uninhibited reactions.
[A:112]

201 If more than one substrate is involved, the reaction is studied by measuring a set of initial velocities for a range of substrate A concentrations, whilst concentrations of other substrates are kept constant. An inhibitor can thus be one of the following types—competitive, noncompetitive, and uncompetitive with respect to this substrate. The character of the inhibition can well be different with respect to substrate B.
[D:115]

202 These alcohols are harmful mainly because they are oxidized to produce very toxic products. Ethanol will competitively inhibit alcohol dehydrogenase and thus reduce the rate of formation of these toxic products. In consequence, a greater proportion of the (less harmful) methanol or glycol is excreted unchanged.
[F:119]

## D Allostericity

203 The mechanism is not of the Michaelis–Menten type because the $v$ against $[A]$ plot is not hyperbolic. The sigmoidal shape could arise from a cooperative effect. If the enzyme molecule contains more than one site to which the substrate may attach, not all these sites being necessarily catalytic, it is possible that some enzyme–substrate complexes might be more powerful catalysts than the enzyme itself, and that the enzyme activity might increase greatly when the concentration of such complexes is appreciable.
[B:300]

204 Allosteric regulation of the enzyme activity is frequently found in enzymes where the molecule is made up of subunits.
[B:299]

205 In positive cooperativity, a substrate molecule, suitably attached to the enzyme molecule, enhances activity at an unoccupied catalytic site,

whereas in negative cooperativity, activity at the catalytic site is reduced.
[B:299]

206     Attachment of a molecule at an allosteric site can slightly alter the geometry of the whole structure, the effect being transmitted to the catalytic site(s) to increase or reduce catalytic activity.
[F:123]

207     Yes. When ATP is a substrate in a reaction, present in a pathway that results in the net production of ATP, a sufficiently high ATP concentration may act to 'switch off' the reaction.
[D:120]

208     No. Allosteric effectors can be products or they can be other activating or inhibiting molecules.
[G:221]

209     They may appear at the 'end' or be required at the 'beginning' of a metabolic sequence. If the former, they may act on an enzyme early in the sequence to switch off its activity when the concentration of the effector exceeds its physiological optimum (negative feedback). If the latter, they may stimulate a reaction later on in the sequence (positive feedforward).
[D:121]

210     The term key enzyme is often used for rate-controlling enzymes (inherently the least active) in a biochemical pathway. These are often present just before or just after a branch in the pathway, and are often subject to regulation by allosteric effectors.
[F:266]

211     When the concentration of citrate rises to an undesirable level in a cell, the citrate will allosterically inhibit phosphofructokinase, the effect being to substantially reduce the rate of glycolysis.
[F:267]

212     Citrate, when its intracellular concentration is high, will increase the activity of acetyl CoA carboxylase and thus stimulate fatty acid synthesis.
[F:544]

## E  Clinical enzymology

213  Suitable isoenzyme examples are:
(a) cytoplasmic and mitochondrial malate dehydrogenases,
(b) acid phosphatases of erythrocytes and of prostatic cells,
(c) lactate dehydrogenase enzymes $LDH_1$ (heart) to $LDH_5$ (liver and skeletal muscle).
[A:134, 137, 1278]

214  The usual methods involve either electrophoresis or appropriate kinetic studies.
[A:135, 202]

215  Reaction velocity measurements obtained under appropriate standard conditions with one substrate (usually the physiological substrate) may be compared with measurements obtained
(a) with a second substrate,
(b) with a special inhibitor,
(c) in the presence of a specific denaturing substance (ethanol is sometimes used),
(d) at an elevated temperature.
One isoenzyme might be much more sensitive than the other to the changed conditions.
[A:136]

216  This is because anticoagulants, used in the collection of plasma, inhibit or inactivate many enzymes.

217  Lactate dehydrogenase [especially the LDH 1 and 2 (heart specific) isoenzymes], aspartate transaminase and creatine phosphokinase determinations can be helpful. Enzyme determinations may be of particular importance (a) if ECG evidence is unsatisfactory and (b) to confirm a diagnosis in a patient who has recovered from an episode (heart specific LDH may show elevated values for up to 12 days following an infarct).
[I:345]

218  A marked rise in serum creatine phosphokinase levels occurs in the Duchenne type of muscular dystrophy (with abnormal values in female carriers) and in some other myopathies.
[I:346]

219 Erythrocytes possess high acid phosphatase activity, which escapes in clotting. Also, prostatic acid phosphatase isoenzyme is more sensitive than red cell acid phosphatase to alcohol. Determinations should be designed to isolate the prostatic acid phosphatase activity and exclude the activity due to erythrocytes. The determinations are of value in the diagnosis of carcinoma of the prostate (values are particularly high in metastasizing prostatic carcinoma) and in following the patient's response to treatment.
[I:343]

220 Examples are:
(a) rickets and osteomalacia,
(b) healing of fractures,
(c) Paget's disease,
(d) malignant disease with an osteoblastic response.
[I:343]

221 In general, damaged tissue releases increased quantities of enzymes into the blood. However, where enzymes, enzyme precursors, or enzyme activators are being synthesized, it is also possible that necrosis may result in reduced serum enzyme activities. Thus it is common for liver cell damage to produce increased serum levels of aspartate and alanine transaminases, but to produce reduced serum cholinesterase activity.
[I:335]

222 Elevated serum concentrations of several enzymes (e.g., alkaline phosphatase, transaminases, isocitrate dehydrogenase) can indicate symptomless hepatocellular damage.
[I:290, 297]

223 Alkaline phosphatase activity is raised in cholestasis and amylase activity in acute pancreatitis.
[I:295, 341]

# SOME IMPORTANT TYPES OF SUBSTRATE MOLECULE

## A Carbohydrates

224 These are easily distinguished in open chain structures, an aldose possessing a terminal –CHO group and a ketose possessing a $>$CO group in the chain. In cyclic structures, one ring

carbon atom will be linked to 2 oxygen atoms
and to a carbon atom. If the remaining valency is
linked to a hydrogen atom the structure is an
aldose, if linked to another carbon atom it is a
ketose.
[F:257]

225  If simple carbohydrates are defined in terms of
aldehyde or ketone derivatives of diols or other
polyhydric alcohols, then the simplest examples
must be the trioses glyceraldehyde (glycerose)
$HOCH_2.CHOH.CHO$  and  dihydroxyacetone
$HOCH_2.CO.CH_2OH$.
[F:257]

226  Monosaccharides are the simplest sugars, they
are carbohydrates that cannot be hydrolysed to
give simpler carbohydrates. Examples are glucose and galactose.
[G:12]

227  Pentoses and hexoses are both monosaccharides.
Pentoses contain a chain of 5 carbon atoms and
hexoses a chain of 6 carbon atoms. These chains
may be part of a cyclic structure. Examples are:

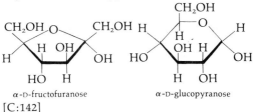

α-D-ribose
(pentose)

α-D-glucose
(hexose)

[C:144]

228  The open chain structures of hexoses may cyclize
to give either 5- or 6-membered, oxygen-
containing, rings. These structures are called
furanose and pyranose ring structures respectively. The stability of a particular ring structure
is dependent on the sugar concerned. Compare
the α-D-fructose structure (furanose) with the α-
D-glucose structure (pyranose) below.

α-D-fructofuranose

α-D-glucopyranose

[C:142]

105

229 A disaccharide molecule contains 2 monosaccharide residues united by a glycosidic linkage (with the elimination of $H_2O$), e.g. maltose

contains 2 glucose molecules linked with the elimination of a water molecule. Polysaccharides contain many monosaccharide residues linked by glycosidic linkages. An example is glycogen.
[C:147]

230 A D-carbohydrate molecule is always the mirror image of the corresponding L-carbohydrate molecule. Note, however, that a D-carbohydrate might give (+) *or* (−) rotation to the plane of polarization of plane-polarized light.
[C:142]

231 $\alpha$- and $\beta$- anomers of D-glucose are stereoisomers that only differ in the configuration at carbon 1. An aqueous solution of either $\alpha$-D-glucose or $\beta$-D-glucose gradually undergoes mutarotation to give an equilibrium mixture of $\alpha$- and $\beta$- structures in solution.
[F:258]

232 Readily oxidized sugar molecules reduce Benedict's reagent and similar reagents. These reducing sugars possess aldehyde or hydroxyketone groups or their cyclized variants. In non-reducing sugars these groups are modified by substitution. Thus all (unsubstituted) monosaccharides are reducing sugars, whilst sucrose is a notable non-reducing disaccharide.
[C:147]

233 Classifications are as follows:
  (a) amylose, an unbranched polysaccharide (an unbranched polyglucose);
  (b) arabinose, an aldopentose;
  (c) lactose, a (reducing) disaccharide;
  (d) fructose, a ketohexose;
  (e) glycogen, a branched polysaccharide (a branched polyglucose);
  (f) mannose, an aldohexose.
  [C:144]

## B Triglycerides

234 Triglycerides are esters of glycerol.
[C:188]

235 The long chain triglycerides are lipophilic.
[C:16]

236 Triglycerides are hydrolysed by various lipase enzymes, the products sometimes being mono- or diglycerides. Chemically they may be hydrolysed by boiling with acids or (more commonly) with alkalis. The products of complete hydrolysis (enzyme or chemical) are glycerol and fatty acids. Fatty acids will separate as salts (soaps) from concentrated solutions under alkaline conditions.
[D:351, 380]

237 A saturated straight aliphatic chain $CH_3(CH_2)_{16}-$ is attached to the carboxyl group in stearic acid. A similar chain containing one double bond $CH_3(CH_2)_7CH=CH(CH_2)_7-$ is present in oleic acid.
[F:384]

238 The mechanism of synthesis of fatty acids, starting with acetyl CoA, elongates the chain in 2-carbon steps. This necessarily produces acids with an even number of carbon atoms (e.g. $C_{15}H_{31}.COOH$). $\beta$-Oxidation of fatty acids also occurs in 2-carbon steps and operating on an acid with an odd number of carbon atoms will leave a 3-carbon residue. Acid $C_{16}H_{33}.COOH$ is not known to play any part in human metabolic processes.
[C:186]

239 These acids are known as polyunsaturated fatty acids.
[C:187]

240 In the $\Delta$ double bond nomenclature, numbering starts with the carboxyl C atom, whilst in the $\omega$ nomenclature, it starts with the terminal $CH_3$ group.
[C:186]

241 Chain lengthening or shortening in fatty acids occurs at the carboxyl end of the molecule and so the number of an individual carbon atom does

107

not alter if numbering starts with the terminal $CH_3$.
[F:395]

242 A saturated fatty acid $C_{12}H_{26}O_2$ has greater metabolically available energy per molecule than a $C_{12}$ carbohydrate ($C_{12}H_{22}O_{11}$, if a disaccharide). Humans can metabolize both to $CO_2$ and $H_2O$, but the carbohydrate can be considered as a structure that is already extensively oxidized. The fatty acid also has greater metabolic energy per gram than the carbohydrate.
[C:570]

## C  Nucleotides

243 A nucleoside consists of a pyrimidine molecule linked from its 1-N nitrogen atom, or a purine molecule linked from its 9-N nitrogen atom, to a sugar molecule (usually D-ribose or 2-deoxy-D-ribose) at the 1-C atom. A nucleotide is a phosphate ester of a nucleoside.
[C:325]

244 Nucleoside phosphates may be esters of orthophosphoric acid (as in AMP) or of more complex phosphoric acids (as in ADP and ATP).
[C:327]

245 Pyrimidine consists of a single 6-membered nitrogen-containing aromatic ring, with N atoms at the 1- and 3-positions. Purine contains a pyrimidine ring and a 5-membered (imidazole) ring, the two rings possessing 2 adjacent carbon atoms in common. The imidazole ring contains N atoms at positions 7 and 9 of the purine structure.
[C:323]

246 The structures refer to the following:
(a) adenine A, a purine,
(b) uracil U, a pyrimidine,
(c) cytosine C, a pyrimidine,
(d) guanine G, a purine.
[C:324]

247 A base of similar importance is thymine T, the 5-methyl derivative of uracil (structure (b)).
[C:324]

248 Abbreviations A, U, C, G and T are used to refer not only to the bases, but also to the corresponding nucleosides, nucleotides, and nucleotide residues in polynucleotide and nucleic acid chains.
[F:513]

249 The bonds linking bases to ribose structures join atom 1-N (in pyrimidines) or 9-N (in purines) to ribose atom 1-C. Two stereochemical possibilities exist at this carbon atom ($\alpha$ and $\beta$). In the important nucleosides the N–C bond has the $\beta$-orientation.
[F:512]

250 The products are:
(a) sugar–phosphate and base (as hydrochloride),
(b) nucleoside and sodium phosphate.

251 Structures of AMP and dAMP differ at the 2-C position of the ribose ring:

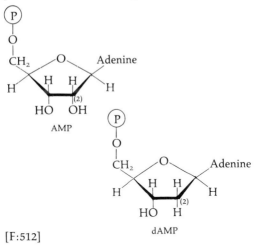

[F:512]

252 Polyphosphoric acid structures are strongly acidic and a considerable proportion (80%) of ATP in solution at pH 7 will be present in the fully ionized $ATP^{4-}$ form.
[A:190]

## D Other significant substrate molecules

253 The NADH structure is made up of 2 base-substituted ribose monophosphates linked by the

elimination of water between the 2 phosphate groups. In NADPH an additional phosphate has been added at position 2 of the adenosine ribose ring. At pH 7 and above, the phosphate groups give these molecules negative charges, NADPH carrying a larger negative charge than NADH. Conventionally, these negative charges are ignored when discussing dehydrogenation reactions.
[F:244]

254 Stoichiometrically, the nicotinamide ring accepts an electron and a hydrogen atom. The process may be written as:

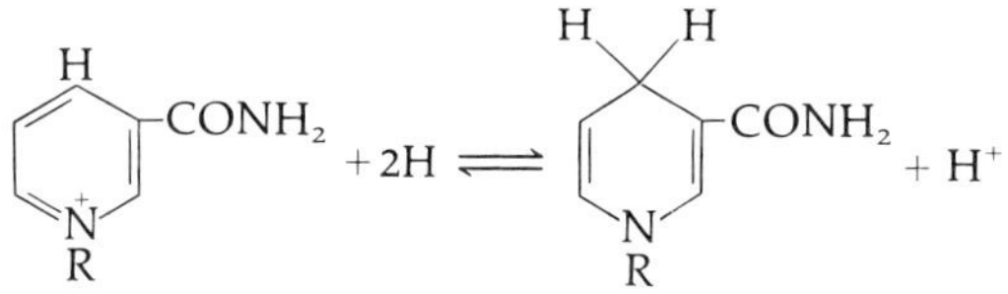

or conventionally

$NAD^+ + 2H \rightleftharpoons NADH + H^+$

$NADP^+ + 2H \rightleftharpoons NADPH + H^+$

[F:245]

255 Three examples of phosphoric acid derivatives with high group transfer potential are:

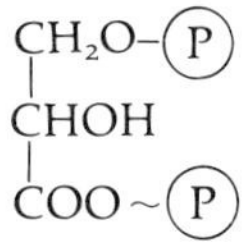

1,3-Diphosphoglycerate
(an acid anhydride)

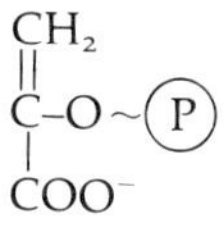

Phosphoenolpyruvate
(an unsaturated ester)

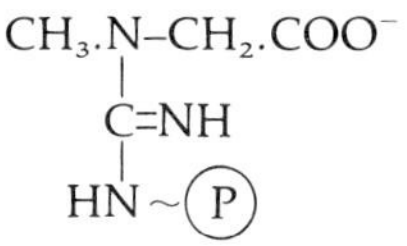

Creatine phosphate
(an imide)

[F:242]

256 The triphosphate part of the ATP structure is extensively ionized and most molecules carry a charge of $-4$ at pH 7. Charge separation makes energy available when the terminal phosphate is transferred. The reaction mechanism involves

$ATP^{4-}/Mg^{2+}$ complexes with the kinase but the separation of the negative charges is still the main energy source.
[A:189]

257 Coenzyme A consists of an ADP molecule with its terminal phosphate linked to a chain of atoms (pantetheine) that terminates in an –SH group. The main feature is the –SH group, which reacts to form thiol esters.
[F:247]

258 The flavin-containing group in these aerobic dehydrogenases may be riboflavin phosphate (FMN) or (ribo)flavin adenine dinucleotide (FAD). Usually 2 molecules of flavin nucleotide, together with metal ions, are present per enzyme molecule. Riboflavin (vitamin $B_2$) comprises a fused tricyclic N-containing aromatic base linked to ribitol. In FAD the ribitol is then linked to the terminal phosphate of an ADP molecule. The mechanism of the dehydrogenating action is complex, but the N-ring structure can reversibly accept 2 H atoms.
[F:245]

259 Coenzyme Q is a quinone of structure

R = an unsaturated hydrocarbon (polyisoprenoid) side chain

Quinones of this type can accept electrons and transfer them to cytochrome b. Vitamin K is also a quinone (2-methylnaphthaquinone); the natural vitamin variants also possess isoprenoid side chains at position 3.
[F:313, 249]

260 Pyridoxal phosphate and pyridoxamine phosphate both possess a pyridine ring with substituents on four adjacent carbon atoms. The difference is in the substituent in position 4 (*para* to the ring N). The aldehyde group –CHO occupies this position in pyridoxal phosphate and it is converted into $-CH_2NH_2$ in pyridoxamine phosphate.
[F:409]

# GLYCOLYSIS AND FATTY ACID OXIDATION

## A  Glycolysis

261  In aerobic glycolysis, a glucose molecule is converted, by several steps, into 2 pyruvic acid molecules. This indirectly requires oxygen as:

$$C_6H_{12}O_6 = 2C_3H_4O_3 + 4H.$$

In anaerobic glycolysis, the 2 pyruvic acid molecules are reduced to give 2 lactic acid molecules. In the overall process from glucose, no oxygen is required because

$$C_6H_{12}O_6 = 2C_3H_6O_3.$$

[F:269, 352]

262  Two molecules of the hydrogen acceptor $NAD^+$ are converted into $NADH + H^+$ when one glucose molecule is converted into 2 pyruvic acid molecules. Each pyruvic acid molecule then utilizes an $NADH + H^+$ for lactic acid formation resulting in no net change in the quantity of $NAD^+$.
[F:263, 352]

263  The reaction is the scission of fructose 1,6-diphosphate to give glyceraldehyde phosphate and dihydroxyacetone phosphate. This is a reverse aldol condensation aided by charge separation of the two negatively charged phosphate groups. The enzyme is aldolase.
[F:262]

264  The conversion of glucose into glucose 6-phosphate and of fructose 6-phosphate into fructose 1,6-diphosphate both require ATP, whilst ATP is formed from ADP in the conversion of 1,3-diphosphoglycerate into 3-phosphoglycerate and of phosphoenolpyruvate into pyruvate. The net production is $(-2 + 2 \times 2) = 2$ ATP molecules.
[F:260, 265]

265  Glycolysis occurs in the cytosol.
[F:260]

266  As erythrocytes contain no mitochondria, these cells are dependent on glycolysis for their supply

of ATP. In rapidly contracting skeletal muscle, inadequate oxygen supply will lead to anaerobic glycolysis and this to lactate formation.
[C:161]

267 Eight molecules of ATP are produced.
[F:265]

268 The 3 irreversible steps in the glycolytic pathway itself are:

glucose → glucose 6-phosphate,

fructose 6-phosphate → fructose 1,6-diphosphate,

phosphoenol pyruvate → pyruvate.

[C:165]

269 Glucokinase is a specific hexokinase isoenzyme of high $K_m$, present in mammalian liver. It can increase the rate of glucose 6-phosphate production if glucose concentration rises, whilst hexokinases in other tissues are already saturated with glucose.
[C:163, 180]

270 The metabolism of fructose proceeds by the formation of fructose 1-phosphate, which is split by an aldolase to give dihydroxyacetone phosphate and glyceraldehyde. The latter is converted by ATP into glyceraldehyde phosphate, and metabolism proceeds along the later part of the glycolytic pathway.
[F:181]

## B  Fatty acid oxidation

271 ATP is required only for the initial formation of the CoA derivative of the fatty acid in the cytosol.
[F:387]

272 β-Oxidation occurs in the inner matrix compartments of mitochondria.
[F:388]

273 Carnitine is a derivative of β-hydroxybutyric acid, possessing a $(CH_3)_3\overset{+}{N}-$ group attached to its terminal methyl group. This compound transports fatty acyl groups through the inner mitochondrial membrane, in the form of carnitine

113

esters, and then transfers them to intramitochondrial CoA molecules.
[F:388]

274 The term $\beta$-oxidation is used because, in the process, oxidation occurs at the $\beta$-carbon atom to form a ketone:

$$R{-}\underset{\beta}{CH_2}{-}\underset{\alpha}{CH_2}{-}COS(CoA) {-}{-}{-}\rightarrow$$

$$R{-}\underset{\beta}{CO}{-}\underset{\alpha}{CH_2}{-}COS(CoA).$$

[C:201]

275 The coenzymes required are FAD, $NAD^+$ and CoA.
[F:389]

276 If $\beta$-oxidation has been occurring for a sufficient time, then the following sequence will be present: $C_{13}H_{27}.COS(CoA)$, $C_{11}H_{23}.COS(CoA)$, $C_9H_{19}.COS(CoA)$, $-{-}{-}C_3H_7.COS(CoA)$, and $CH_3.COS(CoA)$.
[F:390]

277 The product will be acetyl CoA.
[F:389]

278 The number of CoA molecules required will be $(n+1)$.
[B:428]

279 The number of molecules reduced are:

FAD (7 molecules) $\rightarrow$ 7 $FADH_2 \equiv 14$ ATP

$NAD^+$ (7 molecules) $\rightarrow$ 7 $NADH \equiv 21$ ATP

[F:391]

280 The acid produced will be phenylacetic acid. It will be excreted in the urine as its glycine conjugate, phenylaceturic acid.
[F:387]

# CITRIC ACID CYCLE, ELECTRON TRANSPORT AND OXIDATIVE PHOSPHORYLATION

## A Citric acid cycle

281 Pyruvate oxidation occurs in mitochondria. Most tricarboxylic acid cycle enzymes are in the lumen

of mitochondria, while others are attached to the inner membrane.
[F:283]

282　Acetyl CoA is produced not only from pyruvate but also by fatty acid oxidation. The irreversibility of the pyruvate dehydrogenase-catalysed reaction prevents the synthesis of glucose from fatty acids in mammals. No net synthesis from acetyl CoA is possible by traversing the cycle and no alternative pathway exists in mammals.
[F:394]

283　Pyruvate first reacts with the thiazole ring of TPP with the elimination of $CO_2$ and the formation of $CH_3.CHOH–TPP$. This TPP derivative reacts with the disulphide bridge in lipoate $S.CH_2.CH_2.CHR.S$ [where $R = -(CH_2)_4.COO^-$] to form a mono-acetyl derivative $HS.CH_2.CH_2.CHR.S.COCH_3$, which then transfers its acetyl group to a CoA molecule leaving a reduced lipoate. Reaction with $NAD^+$ reconverts the latter into the oxidized (disulphide) form.
[C:166]

284　The formation of succinyl CoA from $\alpha$-oxoglutarate is analogous to the formation of acetyl CoA from pyruvate.
[F:294]

285　Acetyl CoA reacts with oxaloacetate to form citrate in the presence of citrate synthase.
[F:284]

286　The 2 steps that produce $CO_2$ are
(a) the conversion of isocitrate into $\alpha$-oxoglutarate;
(b) the conversion of $\alpha$-oxoglutarate into succinyl CoA.
[F:285]

287　The formation of succinate from succinyl CoA in the presence of succinyl thiokinase occurs with the concomitant production of GTP from GDP.
[F:286]

288 Consider the reaction sequence:

$$
\begin{array}{c}
R \\
| \\
CH_2 \\
| \\
CH_2 \\
| \\
COO^-
\end{array}
\underset{FADH_2}{\overset{FAD}{\rightleftharpoons}}
\begin{array}{c}
R \\
| \\
CH \\
|| \\
CH \\
| \\
COO^-
\end{array}
\underset{-H_2O}{\overset{+H_2O}{\rightleftharpoons}}
\begin{array}{c}
R \\
| \\
CHOH \\
| \\
CH_2 \\
| \\
COO^-
\end{array}
\underset{NADH}{\overset{NAD^+}{\rightleftharpoons}}
\begin{array}{c}
R \\
| \\
CO \\
| \\
CH_2 \\
| \\
COO^-
\end{array}
$$

If R is an alkyl group this represents a sequence in $\beta$-oxidation and if R is a carboxylate ion it represents the sequence from succinate to oxalo-acetate in the citric acid cycle.
[F:287, 389]

289 The oxidation of a pyruvate ion gives 4 NADH + 1 $FADH_2$ + 1 GTP, equivalent to 15 ATP.
[F:283, 288]

290 Various intermediates in the cycle can be pro-duced or removed by other (non-cycle) reactions. Some are starting points for the synthesis of other important metabolites; oxaloacetate and $\alpha$-oxoglutarate are available for the transamination of amino acids; and oxaloacetate is a key inter-mediate in gluconeogenesis.
[F:299]

## B Electron transport and oxidative phosphorylation

291 Electron transport and oxidative phosphorylation occur at the inner mitochondrial membrane.
[F:307]

292 The likely sequence of electron flow in the re-spiratory chain, starting with NADH, can be summarized as follows:

NADH → (Flavoprotein–FeS complex) →

Coenzyme Q

→ (Cyt b–FeS complex) → Cyt $c_1$ →

Cyt c → Cyt a

→ Cyt $a_3$ → $O_2$.

[F:311]

293 The reaction that triggers the production of the first ATP molecule lies between NADH and CoQ in the above sequence, the second ATP

molecule is produced as electrons move from
Cyt b to Cyt c$_1$ and the third as they move from
Cyt a to oxygen.
[C:133]

294 The sequence from succinate is

$$FADH_2 \rightarrow CoQ \rightarrow Cyt\, b \rightarrow Cyt\, c_1 \rightarrow Cyt\, c$$

$$\rightarrow Cyt\, a \rightarrow Cyt\, a_3 \rightarrow O_2.$$

Here again, the $FADH_2$ and Cyt b are part of
FeS complexes. The number of ATP molecules
produced (per oxygen atom reduced) is 2, these
correspond to the second and third ATP
molecules produced in the sequence from
NADH.
[C:113]

295 Both cyanide and sulphide ions inhibit cytochrome oxidase $(a + a_3)$ and thus block the operation of the respiratory chain oxidation.
[C:134]

296 It is possible for the 2 processes to be uncoupled
and for electron transport to proceed without
oxidative phosphorylation.
[C:134]

297 In the coupled system, the mitochondrial ratio
[ATP]/[ADP], which will regulate the rate at
which oxidative phosphorylation occurs, will also
regulate the rate of electron transport. Low ADP
levels stimulate both processes.
[C:132]

298 2,4-Dinitrophenol, gramicidin and valinomycin
are examples of uncoupling agents.
[C:134]

299 Dinitrophenols were once used as slimming
drugs, but high toxicity made their use
undesirable.
[D:206]

300 Briefly, electron transport processes starting with
NADH or $FADH_2$ effect the uptake of protons
from the matrix and their secretion to the outer
surface of the inner mitochondrial membrane. In
order to accomplish this charge separation, the
components of the electron transport chain must

be arranged in an ordered fashion in the membrane. Membrane ATPase is also taken to be specifically oriented. This mitochondrial ATPase is located in isolated regions within the membrane and, according to the hypothesis, the proton gradient arising from the respiratory chain sequence provides a source of energy for reversing the highly exergonic ATPase-catalysed hydrolysis of ATP. Reduction of the negative charges on $ADP^{3-}$ and $HPO_4^{2-}$ ions will be involved in the course of transference of protons back to the mitochondrial matrix.
[C:134]

# THE FED STATE

## A  Glycogenesis

301  The linkages are $\alpha$-1,4-glucosidic linkages. Water has been eliminated between the OH group at position 1 of one glucose residue and the OH group at position 4 of the next. As the orientation of the OH at position 1 is $\alpha$, and is similar to that at position 4, the linkage through an oxygen atom is described as $\alpha$-1,4.
[F:357]

302  This second type of linkage, described as an $\alpha$-1,6-glucosidic linkage, is formed by the elimination of water between the OH group at position 1 of an $\alpha$-D-glucose residue (terminating a short chain) and the OH group at position 6 of an $\alpha$-D-glucose residue in a chain. This produces a branch in the latter chain.
[F:357]

303  Small quantities of glycogen are present in many tissues, but the main sites of synthesis and storage are in liver and striated muscle.
[C:167]

304  Glycogen synthesis and storage take place in the cytosol.
[F:358]

305  Glycogen is not transported, but is broken down in the storage cells.
[C:167]

306 There is no pathway difference, but several enzymes differ. Thus the conversion of glucose into glucose 6-phosphate, catalysed by hexokinase in muscle and other tissues, is catalysed by glucokinase in liver under fed state conditions when blood glucose concentration is raised.
[C:169]

307 The synthetic route, summarized briefly, is

$$\text{glucose 6-P} \rightleftharpoons \text{glucose 1-P} \xrightarrow{\text{UTP}} \text{UDP-glucose}$$

$$\xrightarrow[\text{position 4 of a (glucose)}_n \text{ chain}]{\text{this glucose residue adds to}} \text{(glucose)}_{n+1}$$

The enzymes are respectively phosphoglucomutase, UDPG pyrophosphorylase and glycogen synthetase.
[F:364]

308 Action of the branching enzyme produces the $\alpha$-1,6 links. A chain of about 5 residues is removed from the extending end of a chain and this fragment is reattached further down the chain by an $\alpha$-1,6 linkage.
[F:365]

309 Insulin indirectly stimulates glycogen synthetase in the fed state, whilst adrenaline (epinephrine) and glucagon indirectly inhibit this enzyme in the fasting state.
[F:367]

310 Glycogen storage disease type IV (amylopectinosis; Andersen's disease) is caused by a deficient branching enzyme, which leads to an accumulation of abnormal liver glycogen with long straight chains. This rare condition is associated with fasting hypoglycaemia and progressive cirrhosis of the liver.
[F:375]

## B Lipogenesis

311 Fatty acid synthesis occurs in the cytosol.
[F:395]

312 The complex is an associated set of 7 cytoplasmic enzymes that catalyse fatty acid synthesis.
[F:400]

313 Acetyl CoA cannot cross the mitochondrial membrane, but it condenses with oxaloacetate to form citrate, which can pass into the cytosol. The citrate is broken down by a process requiring both coenzyme A and ATP to give oxaloacetate and acetyl groups (as acetyl CoA).
[F:401]

314 Acetyl CoA is first converted into the more reactive malonyl CoA, the carboxylation reaction requiring biotin and ATP as well as $CO_2$. The acetyl and malonyl groups are then transferred to acyl carrier proteins (ACP)SH to form acetyl-S(ACP) and malonyl-S(ACP) respectively. In the presence of a condensing enzyme, these react with the elimination of $CO_2$ and one acyl carrier protein to give acetoacetyl-S(ACP).
[F:395]

315 Still attached to its acyl carrier protein, the acetoacetyl group is reduced by NADPH to the $\beta$-hydroxybutyryl group. Dehydration follows, introducing a double bond. Reduction of the latter by NADPH gives rise to butyryl-S(ACP).
[F:399]

316 Butyryl-S(ACP) reacts with malonyl-S(ACP) and the cycle is repeated, the ACP derivatives being firmly bound to the enzyme complex. The product in man is palmitic acid.
[F:400]

317 In these and other tissues, $\alpha$-glycerophosphate is formed in the cytosol by reduction of dihydroxyacetone phosphate (from the glycolytic pathway) with NADH.
[C:215]

318 $\alpha$-Glycerophosphate can be formed in the liver, which contains glycerol kinase, by reaction of glycerol with ATP. Adipose cells and intestinal mucosa cells lack this enzyme.
[C:215]

319 $\alpha$-Glycerophosphate reacts in turn with 2 fatty acyl CoA derivatives to form a phosphatidic acid. This intermediate is required for both triglyceride

and phospholipid synthesis. Action of a phosphatase, followed by reaction with another fatty acyl CoA molecule gives triglyceride. Liver and adipose tissue are the main synthesizers, but other tissues also synthesize triglyceride.
[C:215]

320 Limited extension of chain length beyond the $C_{15}H_{31}$ alkyl chain of palmitic acid is effected by special microsomal enzymes, and also possibly by mitochondrial enzymes, acting on palmitoyl CoA. The more important microsomal process uses malonyl CoA as acetyl donor and NADPH as the source of reducing power.
[C:208]

## C  The pentose phosphate pathway

321 This sequence of reactions occurs in the cytosol.
[C:173]

322 Enzymes in the pathway are present in large quantities in liver and adipose tissue and the pathway is also of importance in lactating mammary tissue, adrenal cortex, testes and other tissues concerned with steroid or fatty acid synthesis.
[C:173]

323 The pathway acts as
   (a) a source of NADPH required in the synthesis of fatty acids, steroids and glutathione,
   (b) a source of ribose 5-phosphate for nucleotide synthesis,
   (c) a means for converting $C_x$ monosaccharide phosphates into $C_y$ monosaccharide phosphates (e.g. the interconversion of pentose monophosphates and hexose monophosphates).
[F:333]

324 Glucose 6-phosphate dehydrogenase effects the oxidation of glucose 6-phosphate with $NADP^+$, the lactone produced being hydrolysed (by a lactonase) to give 6-phosphogluconate. Then a second $NADP^+$ molecule (in the presence of 6-phosphogluconate dehydrogenase and $Mg^{2+}$) oxidizes the gluconate to give a pentose phosphate with the elimination of $CO_2$. Two molecules of

NADPH are produced by these processes.

$$G\ 6\text{–}P \xrightarrow[+H_2O]{-2H} \textcircled{P}\text{–}OCH_2.(CHOH)_4.\overline{\underline{COO}}H$$

$$\xrightarrow[-CO_2]{-2H} \textcircled{P}\text{–}OCH_2.(CHOH)_2.CO.CH_2OH$$

[F:334]

325 Transketolases catalyse the transfer of the 2-carbon unit $HOCH_2.CO-$ (together with a hydrogen atom) from one monosaccharide phosphate to another. This process requires the presence of thiamine pyrophosphate (TPP) and $Mg^{2+}$ ions.
[F:335, 340]

## D Unsaturated fatty acids

326 Oleic, linoleic, linolenic, and arachidonic acids have respectively 1, 2, 3 and 4 C=C double bonds, all of *cis* configuration in the naturally occurring isomers. Only the all *cis* isomers are the active forms of the essential unsaturated fatty acids.
[C:209, 212]

327 The CoA derivative of stearic acid is acted upon by a mixed function oxidase, of the endoplasmic reticulum, producing the CoA derivative of oleic acid. The reaction is as follows:

$$C_{17}H_{35}.CO\text{–}S(CoA) + NADPH + H^+ + O_2$$
$$\downarrow$$
$$C_{17}H_{33}.CO\text{–}S(CoA) + NADP^+ + 2H_2O.$$

[C:210]

328 There is no reaction in mammals that converts oleic acid into the essential unsaturated fatty acid, linoleic acid. Arachidonic acid can be synthesized from linoleic, but is preferably present in the diet. Linoleic and linolenic acids must be present in the diet.
[C:210]

329 Mammalian cells lack enzymes that can effect the insertion of double bonds between adjacent carbon atoms from carbon $\omega$-1 to carbon $\omega$-7.
[C:210]

330 Essential unsaturated fatty acid deficiency could develop in patients receiving all nourishment parenterally without unsaturated lipid supplement, in patients with severe malabsorptive defects, and in infants fed on skim milk. Any artificial, low lipid diet could lead to this deficiency.
[D:342, 374]

## E  The hypoglycaemic hormone

331 The important substance is D-glucose, which stimulates the secretion of insulin when the plasma glucose concentration is raised. However, in addition to glucose, some amino acids, some hormones (e.g., secretin, pancreozymin, glucagon) and some drugs (e.g. tolbutamide) can stimulate secretion.
[A:557, 345]

332 No. Insulin is required for uptake by muscle, adipose tissue and some other tissues, but many tissues, notably brain, liver and kidney, are freely permeable to glucose.
[A:324]

333 Insulin exerts an effect by stimulating liver glucokinase, phosphofructokinase and pyruvate kinase activities.
[A:324]

334 Insulin stimulates glycogen formation and also carbohydrate oxidation in muscle and liver. It stimulates lipogenesis and the pentose phosphate pathway in adipose cells and in liver. It also has inhibiting effects on gluconeogenesis and on the release of fatty acids from adipose tissue.
[A:324]

335 It has an anabolic effect by facilitating the uptake of some amino acids and, as an indirect consequence of its other activities, the *net* breakdown of protein to provide energy over any extended time interval is usually rendered unnecessary. Insulin also stimulates the uptake of $K^+$ ions by cells.
[A:324, 551]

# THE FASTING STATE

## A  Fasting state hormones

336  The following are hyperglycaemic hormones: adrenaline (from the adrenal medulla), glucagon (from pancreatic $\alpha$- or A-cells), growth hormone (somatotropin, GH) and adrenocorticotropic hormone (ACTH) (from the anterior pituitary), and glucocorticoids (notably, cortisol) (from the adrenal cortex).
[A:352]

337  Adrenaline is a catecholamine. It is the 3,4-dihydroxy derivative of N-methyl-phenylethanolamine $[(HO)_2C_6H_3.CHOH.CH_2.NHCH_3]$. Glucagon, GH and ACTH are polypeptides. Glucocorticoids are steroids containing 21 carbon atoms.
[D:597]

338  In the fasting state, these hormones operate to maintain an adequate blood glucose concentration, the principal purpose of which is to preserve the supply of glucose to the brain.
[I:182]

339  Both adrenaline and glucagon initiate cAMP synthesis and, in consequence, switch on glycogen breakdown and the mobilization of fatty acids from triglyceride, whilst switching off the reverse synthetic processes. They also stimulate gluconeogenesis. Both hormones exert their effects rapidly. Whereas glucagon is more active than adrenaline in liver, the relative activities are reversed in adipose tissue and in skeletal muscle.
[F:549]

340  Glucocorticoids have a catabolic action in muscle and adipose tissue, stimulating the breakdown of protein and triglyceride and inhibiting the uptake of glucose by these tissues. They possess an anabolic action in liver, one consequence of which is increased synthesis of gluconeogenic enzymes.
[C:490]

341  The activities of these peptide hormones are complex. Both are hyperglycaemic and both inhibit glucose uptake by various tissues. Possibly by

an indirect, and slowly acting, effect on lipase activity, GH stimulates the release of fatty acids from adipose tissue triglyceride. ACTH stimulates the synthesis and subsequent release of corticosteroids by adrenals and thus indirectly influences fasting state metabolism. It also has a more direct, and rapid, effect by causing adipose cells to produce cAMP and thus accelerating the release of fatty acids from triglycerides in these cells.
[C:505, 510]

## B  Glycogenolysis

342  Glycogenolysis occurs in the cytosol. Most cells are capable of glycogen metabolism, with liver and muscle being particularly important.
[F:358]

343  Non-reducing terminal glucose residues are removed, one at a time, each producing a glucose 1-phosphate molecule:

Inorganic phosphate + Glucose–(glucose)$_n$

$$\downarrow \text{Phosphorylase}$$

Glucose 1-phosphate + (Glucose)$_n$.

This continues along each chain, ceasing about 3 to 4 residues from a branch point. The resulting polysaccharide is called a limit dextrin.
[F:359]

344  A transferase enzyme cleaves a short terminal segment leaving just a single glucose residue at the branch. The cleaved segment is joined to the non-reducing terminal of a mainstem, short, terminal segment. Single glucose residue branches will then exist on chains that will extend about 6 residues beyond the branch. A glucosidase then removes the single residue stubs. Overall, a debranching process can be written as

$$(\text{Glucose})_m + H_2O \xrightarrow[\text{enzyme}]{\text{'debranching}'}$$
with
$x$ branches

$$\text{Glucose} + (\text{Glucose})_{m-1}$$
with
$x - 1$ branches

[F:360]

345 The principal control is exerted by hormones such as adrenaline and glucagon. These attach to receptor sites on target cells and stimulate the formation of cAMP in the cells. The cAMP induces an enzyme cascade that produces active phosphorylase enzyme.
[F:366]

346 Glucose 1-phosphate is converted into the 6-phosphate which, in liver, is hydrolysed by glucose 6-phosphatase to give glucose. The main function in liver is to maintain an adequate blood [glucose] for cerebral function in the fasting state.
[F:363]

347 Muscle tissue requires large reserves of fuel for ATP production and so combined glucose is conserved. Muscle (and brain) lack glucose 6-phosphatase and, consequently, the only glucose possibly available to the circulation is that liberated from the single residue stubs of partly degraded glycogen.
[C:167]

348 The most common glycogen storage disease is type 1 (von Gierke's disease). This arises from an inborn deficiency of glucose 6-phosphatase.
[F:374]

349 The patient is likely to be a backward child, possessing a distended abdomen, and experiencing fasting hypoglycaemic episodes. The liver will be enlarged and liver cells loaded with glycogen. Blood lactate, pyruvate, fatty acids, triglycerides and ketone bodies will probably be high and the patient acidotic. Liver glucose 6-phosphatase activity will be very low, but other enzyme activities normal.
[I:202]

## C Gluconeogenesis

350 A continual supply of glucose is necessary as a source of energy for tissues, particularly for the nervous system and for erythrocytes. When the supply of carbohydrates is inadequate, then the carbon skeleton of non-carbohydrate molecules must be converted into glucose. The only suitable non-carbohydrate is protein.
[F:345]

351 Liver and kidney are the main centres of gluconeogenesis in mammals. This pathway is not a
metabolic feature of skeletal muscle and is virtually absent from this tissue.
[F:346]

352 Gluconeogenesis occurs in the cytosol except for
the conversion of pyruvate to oxaloacetate which
takes place in mitochondria. As oxaloacetate cannot itself pass through the mitochondrial membrane, it must be reduced to the diffusible malate
before the pathway can proceed in the cytosol.
[F:346]

353 Glycolysis in muscle under relatively anaerobic
conditions will convert blood glucose (and muscle glycogen) to lactate. This lactate is transferred in the circulation to the liver, where gluconeogenesis can reconvert it into glucose. The
latter re-enters the circulation.
[F:352]

354 Hormonal action induces the breakdown of muscle protein to give amino acids and these undergo
transamination reactions with muscle pyruvate to
form alanine. This passes to the liver where it is
reconverted into pyruvate.
[C:251]

355 $\alpha$-Oxoglutarate is converted into oxaloacetate by
citric acid cycle reactions in mitochondria and
pyruvate is converted into oxaloacetate in mitochondria by the process:

$$CH_3.CO.COO^- + CO_2 \xrightarrow[+\text{biotin}]{\text{ATP}+\text{Mg}^{2+}}$$

$$HOOC.CH_2.CO.COO^-$$

The oxaloacetate is reduced to malate, which
passes out into the cytosol and is there reoxidized
to oxaloacetate. The final stage, which produces
phosphoenolpyruvate is

$$HOOC.CH_2.\underset{\underset{COO^-}{|}}{CO} \xrightarrow[-CO_2]{\text{GTP}} CH_2{=}\underset{\underset{COO^-}{|}}{C}{-}O\,\boxed{P}$$

[F:347]

356 The other specifically gluconeogenic reactions

127

are:
(a) hydrolysis of fructose 1,6-diphosphate to fructose 6-phosphate by fructose 1,6-diphosphatase,
(b) hydrolysis of glucose 6-phosphate to glucose by glucose 6-phosphatase.
[F:346]

357 The regulation of gluconeogenesis involves primarily the key enzymes, pyruvate carboxylase, phosphoenolpyruvate carboxykinase and the 2 phosphatases. These are subject to allosteric regulation and their synthesis is repressed by insulin and induced by corticosteroids.
[C:247]

## D  Lipolysis and ketogenesis

358 The primary function of the specialized mesenchymal cells, adipocytes, is the storage of fats. Under hormonal influence lipolysis occurs in the cytoplasm of these cells. Lipolysis also occurs in the lipoprotein lipase region of the capillary wall to break down lipoprotein lipid and in the intestine to effect absorption of dietary fat.
[D:359]

359 Adipose tissue lipase is a system of enzymes that catalyse the stepwise hydrolysis of triglycerides in adipose cells:

Triglyceride → Diglyceride →
+ fatty acid

Monoglyceride → Glycerol
+ fatty acid        + fatty acid
[D:359]

360 Lipoprotein lipase catalyses the hydrolysis of plasma lipoprotein triglyceride. Pancreatic lipase catalyses the partial hydrolysis of dietary triglycerides. Microsomal lipases can hydrolyse triglycerides of medium chain length.
[D:380]

361 Glycerol is transferred to the liver where a kinase can convert it into $\alpha$-glycerophosphate.
[C:215]

362 Free fatty acid is taken in the circulation to heart,

muscles and kidneys for energy production and to the liver for lipid synthesis.
[C:222]

363 Adipose tissue lipolysis is under hormonal control. Thus glucagon, ACTH, thyroid stimulating hormone, adrenaline and noradrenaline exert a direct stimulating effect, whilst cortisol is among the factors having an indirect stimulating effect. Insulin, prostaglandin $E_1$ and nicotinic acid are inhibiting.
[C:224]

364 The traditional term ketone bodies refers to the three entities acetoacetic acid, $\beta$-hydroxybutyric acid and acetone. The two acids will be extensively ionized in plasma.
[C:234]

365 Acetoacetate formation is a mitochondrial process that, in man, occurs almost exclusively in liver cells.
[C:234]

366 Reversal of the thiolase reaction produces acetoacetyl CoA from 2 molecules of acetyl CoA. Then, condensation with another acetyl CoA molecule produces $\beta$-hydroxy-$\beta$-methylglutaryl CoA (HMG–CoA). A lyase splits off an acetyl CoA leaving acetoacetate. The other ketone bodies are formed as follows:

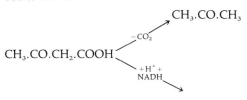

[C:235]

367 Ketone bodies pass from the liver into the circulation. Many extrahepatic tissues contain enzymes that convert acetoacetate (and $\beta$-hydroxybutyrate) into acetoacetyl CoA, which can then be cleaved by the thiolase reaction to give acetyl CoA. The acetyl group is then oxidized by the citric acid cycle and acts as an energy source.
[C:235]

## E Diabetes mellitus

368  The juvenile onset disease is associated with weight loss, in spite of increased appetite, thirst and frequency of urination. An untreated patient will have sustained hyperglycaemia and glycosuria, but will show biochemical features associated with the fasting state.
[D:325]

369  In juvenile onset diabetes, the pancreas is deficient in that the $\beta$-cells are not capable of releasing adequate quantities of active insulin in response to raised plasma glucose concentrations. Appropriate injection of insulin can control the condition.
[C:481]

370  After an overnight fast, blood and urine samples are taken and the patient is given a standard dose of glucose by mouth. Further blood samples are taken at half-hour intervals.
   In normal subjects, plasma glucose concentrations increase in the first hour by 50% or more and then decrease rapidly, the original levels being almost reached within 2 hours. In a diabetic patient, the increase is greater and the return to pretest level is much slower.
[D:327]

371  Increased liberation of free fatty acids occurs and these acids are partly used by tissues. Much is converted into triglyceride in the liver and increasing circulating lipids produce a hyperlipaemia.
[C:481]

372  In juvenile onset diabetes, the liver tends to produce ketone bodies faster than their utilization by muscle and this may produce a ketosis. Of the 3 ketone bodies, only acetoacetic acid is a comparatively strong acid and this is mainly responsible for the development of an acidosis. Respiratory compensation will initially offset this, but if the ketosis persists, an acidosis is likely to develop.
[I:192]

373  Adult onset diabetes is usually milder in character, patients often being elderly and obese.

Polyuria, glucosuria and hyperlipoproteinaemia will be evident.
[B:721]

374 The maturity onset diabetic may show a reduced rate of response to glucose, but the elevated blood glucose concentration may with time lead to relatively high insulin levels. Treatment is usually based on dietary control.
[I:189]

375 The term diabetes refers to any disorder in which polyuria occurs but, apart from this similarity, diabetes mellitus and diabetes insipidus are not related. Polyuria, in the latter condition, arises from a defect in the antidiuretic hormone action. There is no glycosuria, glucose metabolism is normal and the other metabolic abnormalities of diabetes mellitus are absent.
[A:335, 342]

# AMINO ACID METABOLISM

## A  The amino acid pool

376 The amino acid pool consists of the total quantity of free amino acids in the plasma and in other extracellular body fluids.
[B:545]

377 Amino acids are mainly required for protein synthesis, but they are also needed for the synthesis of purines and pyrimidines (required for nucleic acid synthesis), for the production of amines (e.g. catecholamines, the choline and ethanolamine portions of phospholipids, and the glycosamines of connective tissue), for the synthesis of creatine phosphate and of haem. Amino acids are also required for the operation of the urea cycle.
[B:545]

378 Amino acids provide a carbon skeleton for the synthesis of glucose, fatty acids and ketone bodies.
[B:560]

379 The amino acid pool is small, but is involved in rapid turnover, with intake from dietary proteins

and the breakdown of tissue proteins and constant removal for protein and other synthesis.
[F:407]

## B Formation of oxo acids from amino acids

380 The reversible oxidative deamination of L-glutamate is catalysed by glutamate dehydrogenase. The hydrogen acceptor can be either $NAD^+$ or $NADP^+$. The overall reaction is

$$
\begin{array}{c}
\underset{|}{CH_2.CH_2.\underset{|}{CH}.COO^-} \\
\underset{\ }{COOH} \quad \overset{+}{N}H_3 + H_2O
\end{array}
\xrightleftharpoons[\text{NADH} + \text{H}^+]{\text{NAD}^+}
$$

$$
\begin{array}{c}
CH_2.CH_2.\underset{||}{C}.COO^- \\
COOH \quad O + NH_4^+
\end{array}
$$

[F:408]

381 Glutamate dehydrogenase is present in most cells, but particularly in liver cells where activity is present in both cytosol and mitochondria. It is the only amino acid dehydrogenase to be so widely distributed and it participates in the deamination, not only of glutamate itself, but of any amino acid that converts glutarate to glutamate on transamination.
[A:576]

382 Amino acid oxidases are flavoproteins (fp) and, in the reaction, 2H is transferred from the amino acid to the prosthetic group of the flavoprotein, which is then reoxidized by molecular oxygen. The overall reaction is

$$
\begin{array}{c}
R.\underset{|}{CH}.COO^- \\
\underset{\ }{N}H_3 + O_2 + H_2O
\end{array}
\xrightarrow{\text{fp}}
\begin{array}{c}
R.\underset{||}{C}.COO^- \\
O + NH_4^+ + H_2O_2.
\end{array}
$$

L-Amino acid oxidase will not oxidize glycine. This enzyme is found at low activity in liver and kidney. Other tissues do not show appreciable activity.
[A:574]

383 D-Amino acid oxidase is present at high activity in peroxisomes of liver and kidney cells. It oxidizes glycine and L-proline as well as D-amino acids. The stereochemically symmetrical oxo acids that are produced from D-amino acids can

be converted into metabolically useful L-amino acids.
[A:575]

384 Specific dehydrases or desulph-hydrases exist for some amino acids, e.g., serine, homoserine, threonine and cysteine. The reactions are exemplified by that of serine:

$$\underset{\underset{\overset{+}{N}H_3}{|}}{HOCH_2.CH.COO^-} \xrightarrow[+\text{pyridoxal phosphate}]{\text{dehydrase}}$$

$$\underset{\underset{O}{||}}{CH_3.C.COO^-} + NH_4^+.$$

[A:578]

385 Uptake of $NH_3$ or $NH_4^+$ occurs in the conversion of glutaric and aspartic acids to glutamine and asparagine respectively, in the synthesis of carbamoyl phosphate, and in the conversion of oxoglutarate into glutamate (reversed deamination).
[A:579]

386 Transamination is a reversible reaction between $\alpha$-amino acid (1) and $\alpha$-oxo acid (2) to give $\alpha$-oxo acid (1) and $\alpha$-amino acid (2). Pyridoxal phosphate is required as a cosubstrate. It reacts with amino acid (1) to form a Schiff's base, which then undergoes hydrolysis to yield oxo acid (1) and pyridoxamine phosphate. The latter reacts with oxo acid (2), the intermediate Schiff's base hydrolysing to give amino acid (2) and pyridoxal phosphate.
[F:409]

387 Transaminases are present in the mitochondrial matrix. Most tissues possess transaminase activity, but liver, heart, skeletal muscle, kidney, and brain tissue have particularly high activities.
[A:579]

## C Urea cycle and nitrogen balance

388 The main purpose of the urea cycle is to remove ammonia (present mainly as $NH_4^+$ at physiological pH) from the blood. Ammonia has very toxic effects, particularly on the central nervous system and, in consequence its blood concentration is normally very low.
[D:407]

389 The formation of carbamoyl phosphate and its reaction with ornithine to form citrulline takes place in the mitochondrial matrix, but the remainder of the cycle, converting citrulline to ornithine *via* arginosuccinate and arginine, occurs in the cytosol. Carbamoyl phosphate is also synthesized in the cytosol for pyrimidine synthesis.
[B:558]

390 The urea cycle operates in the liver.
[B:555]

391 Ammonia reacts with carbon dioxide and ATP, by a complex reaction involving biotin, to produce carbamoyl phosphate $H_2N.CO–OPO_3^{2-}$. The latter then reacts, transferring the $H_2N.CO–$ group to ornithine, which is a component of the cycle.
[B:557]

392 The enzyme arginase hydrolyses arginine giving urea and ornithine. The urea is transported to the kidneys and excreted in the urine.
[D:274]

393 The following $\alpha$-amino acids are components of the urea cycle: ornithine, citrulline, arginosuccinic acid and arginine. Only arginine can be incorporated into proteins. Aspartic acid is a reactant that converts citrulline to arginosuccinic acid, but it is not strictly a component of the cycle.
[B:556]

394 The reaction between citrulline and aspartic acid requires an ATP molecule, which is split to give AMP and pyrophosphate. In addition, the formation of a molecule of carbamoyl phosphate requires 2 ATP molecules.
[B:556]

395 The man should be in balance and thus should excrete the nitrogen content of his 100 g of protein (about 16 g). This is excreted mainly in urine (as urea, creatinine, uric acid, $NH_4^+$, etc.) with minor (variable) amounts excreted in faeces and sweat.
[B:787]

396 Examples of a positive nitrogen balance would be given by growing children, pregnant women, and patients gaining weight following malnutrition or recovering from an emaciating disease. Negative nitrogen balance might occur in wasting diseases and in old age.
[D:15]

397 Kwashiorkor is a malnutritional disease in children caused by severely reduced protein intake over a long period. It is characterized by lack of tissue development and atrophy of tissues (e.g., pancreas) that require rapid protein synthesis. Oedema and diarrhoea are common.
[B:793]

## D Decarboxylation of amino acids

398 Several amino acids, e.g., ornithine, lysine and tyrosine are decarboxylated by intestinal bacterial enzymes. Some, e.g., histidine, L-dopa (3,4-dihydroxyphenylalanine), and 5-hydroxytryptophan are decarboxylated in body tissues.
[A:587, 610, 642]

399 Decarboxylation of amino acids produces amines, those produced in tissues often possessing neurotransmitter or hormone action.
[A:643, 672, 1058]

400 Examples of amines produced indirectly or directly from amino acids include catecholamines such as noradrenaline and adrenaline, serotonin (5-hydroxytryptamine), and histamine.
[D:419]

401 Amino acid decarboxylations require pyridoxal phosphate. A Schiff's base is formed with weakening of the amino acid $C-COO^-$ bond. Decarboxylation is then followed by hydrolysis to give the amine.
[A:582]

402 Amino acid decarboxylations occur in the cytosol. The aromatic L-amino acid decarboxylase is present in brain, kidney and liver. A distinct histidine decarboxylase is present in a wide range of tissues. L-Dopa is synthesized in brain, in sympathetic nerve endings, in paraganglia and in the adrenal medulla.
[A:643]

403 The biosynthetic pathway for adrenaline is as follows:

Tyrosine $\xrightarrow{\text{hydroxylation}}$ L-Dopa

$\xrightarrow{\text{decarboxylation}}$ Dopamine

$\xrightarrow[\text{hydroxylation}]{\text{side chain}}$

CHOH.CH$_2$.NH$_2$

OH

OH

$\xrightarrow[\text{methionine}]{S\text{-adenosyl}}$

CHOH.CH$_2$.NHCH$_3$

OH

OH

[F:895]

404 Adrenaline is synthesized mainly in the cells of the adrenal medulla, but the pathway must proceed, at least partially, in brain, in paraganglia, and in sympathetic nerve endings.
[D:597]

405 Melatonin is the O-methyl derivative of N-acetylserotonin. After synthesis in the pineal gland, it enters the blood and is taken up by peripheral nerves, sympathetic ganglia, ovaries and the pituitary. It has various activities and may be a regulator of the biological clock.
[D:605]

## E Inborn errors of amino acid metabolism

406 Mutations that alter the nucleotide sequence (genetic code) of the DNA of genes will result in the production of proteins with modified primary structures. Thus, in the case of enzymes, an abnormal amino acid may be present at a catalytic or allosteric site, or an altered 3-dimensional conformation may modify the activity of these sites. Metabolic aberrations will follow.
[I:350]

407 Phenylketonuria, tyrosinosis, tyrosinaemia and alkaptonuria are inborn errors of phenylalanine catabolism. Additionally, tyrosine is converted into melanin pigments. An enzyme defect here produces albinism.
[C:293, 302]

408 Mammalian liver cells contain the phenylalanine hydroxylase complex. Component 1 uses molecular oxygen to convert simultaneously phenylalanine to tyrosine and tetrahydrobiopterin to dihydrobiopterin. Component 2 then uses $NADPH + H^+$ to reduce dihydrobiopterin to give the tetrahydro compound.
[C:266]

409 The defective enzyme in phenylketonuria is component 1 of the phenylalanine hydroxylase complex.
[C:293]

410 The plasma phenylalanine concentration rises and alternative metabolites, phenylpyruvate, phenyllactate and phenylacetate (with its glutamine conjugate) appear in blood and urine. High concentrations of phenylalanine and these metabolites are injurious to the infant brain and mental development is retarded.
[C:293]

411 The ferric chloride urine test (or the Phenistix test) and, if thought necessary, the more reliable Guthrie microbiological assay or the Scriver paper chromatographic method should be carried out in the early weeks of life. If phenylketonuria is detected, then a diet low in phenylalanine can be instituted and continued up to the age of 6 years. At about this age, high phenylalanine concentrations cease to be damaging to the brain.
[I:359]

412 Alkaptonuria is an inborn error of phenylalanine catabolism. Lack of activity of homogentisate oxidase, which opens the substrate's benzene ring, gives rise to urinary excretion of homogentisate. This slowly oxidizes in air to give alkapton, a brownish-black pigment. General pigmentation of connective tissue occurs and an arthritic condition may develop, but the disease is relatively benign.
[C:294]

413 Maple syrup urine disease, a very dangerous condition of infants, arises from the absence (or reduced activity) of the $\alpha$-oxo acid decarboxylase involved in the catabolism of the branched chain aliphatic amino acids, valine, isoleucine and

137

leucine. Transamination converts these amino acids into corresponding $\alpha$-oxo acids. The decarboxylase should then split off $CO_2$ and acyl CoA derivatives remain. Defect in the decarboxylase enzyme leads to a build-up in blood levels of these amino and oxo acids and also formation of the $\alpha$-hydroxy acids (by NADH reduction of the oxo acids).
[C:296]

414  Cystinuria is an inherited metabolic disease in which the excretion not only of cystine, but also of the basic amino acids, lysine, arginine, and ornithine, are greatly increased. It probably arises from a defect in the renal reabsorptive mechanisms for these amino acids. As the solubility of cystine is low, it may crystallize in kidney tubules causing cystine calculi.
[C:295]

415  The autosomal recessive disease, homocystinuria, is caused by a deficiency of cystathionine synthetase. The excess of homocysteine interferes with the maturation of both elastin and collagen.
[C:296]

416  Disorders are known that are associated with defects in each of the urea cycle enzymes. Hyperammonaemia is a common consequence of all these conditions.
[C:275]


# NUCLEOTIDE METABOLISM

## A  Biosynthesis of IMP, AMP and GMP

417  The kinase transfers a pyrophosphate group, rather than a phosphate group, from ATP onto ribose 5-phosphate.
[A:493]

418  The origin of the atoms or groups in IMP is as follows:
(a) the ribose 5-phosphate group is from PRPP,
(b) atoms 3 and 9 are from the amide nitrogens of two glutamine molecules,

(c) atoms 4, 5 and 7 are from glycine,
(d) atom 6 is from carbon dioxide,
(e) atom 1 is from aspartic acid,
(f) atoms 2 and 8 are from formyl-type derivatives of tetrahydrofolate.

The numbers refer to the numbering of the purine ring structure.
[C:334]

419 Tetrahydrofolic acid contains a diazine ring *A* (fused to a second diazine ring *B*) possessing one long side chain $-CH_2NHR$ as indicated.

Formyl derivatives may be produced at positions 5 and 10 and a one-carbon bridge can link these positions. The resulting structures can transfer a one-carbon unit to other molecules. Both $N^5,N^{10}$-methenyl $FH_4$ and $N^{10}$-formyl $FH_4$ are used in purine synthesis.
[A:488]

420 Folic acid antagonists are substances that interfere with the formation of tetrahydrofolate. Methotrexate (amethopterin), a close chemical analogue of folic acid, competitively inhibits the reduction of folate to tetrahydrofolate. It is used to stop or retard the growth of rapidly growing tumours.
[A:487]

421 6-Mercaptopurine and 6-thioguanine are anti-cancer agents concerned with purine metabolism.
[A:507]

422 The conversion of IMP into AMP involves replacement of the OH substituent at position 6 of the enol form of IMP by an $NH_2$ group. This is accomplished by condensation of IMP with aspartate in the presence of a GTP-requiring synthetase to form an intermediate that splits to

give AMP and fumarate.

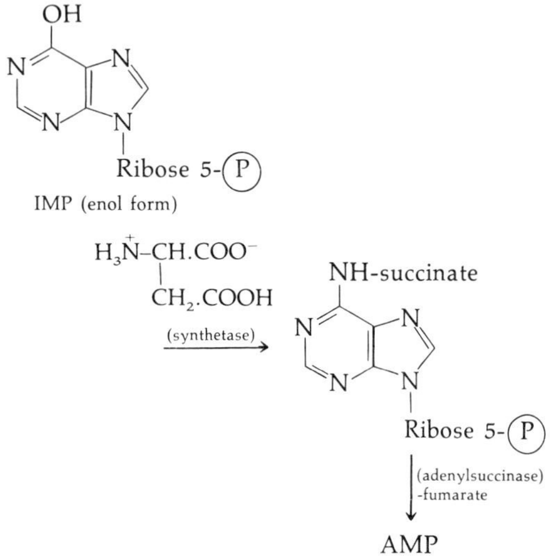

IMP (enol form)

(synthetase)

(adenylsuccinase)
-fumarate

AMP

[C:335]

423 IMP dehydrogenase uses $NAD^+$ to oxidize IMP to xanthosine monophosphate, which is aminated by glutamine in the presence of ATP.

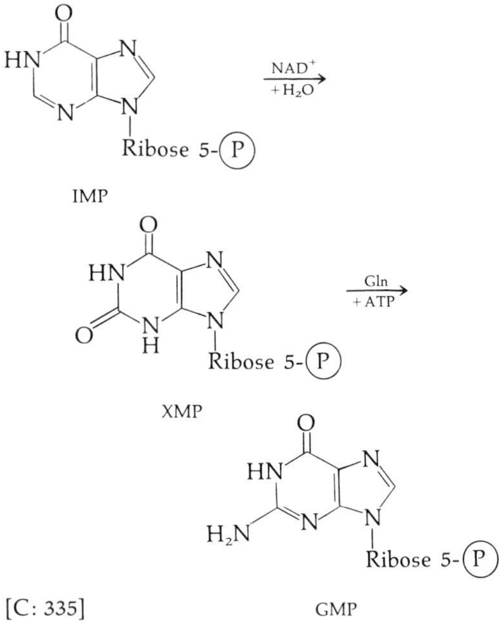

IMP

$\xrightarrow[+ H_2O]{NAD^+}$

XMP

$\xrightarrow[+ ATP]{Gln}$

[C: 335]

GMP

140

## B Biosynthesis of UMP and of ribonucleoside triphosphates

424 In purine biosynthesis, PRPP is converted into 1-amino-5-phosphoribose and the synthesis proceeds from this amino group, building up the purine ring structure. In pyrimidine biosynthesis, the pyrimidine ring structure is assembled before PRPP reacts to form a nucleotide.
[C:334, 341]

425 Aspartic acid provides 3 carbon and one nitrogen atoms.
[F:519]

426 Carbamoyl phosphate condenses with aspartic acid and the product cyclizes to form dihydroorotate. Flavoprotein then oxidizes this to orotate.

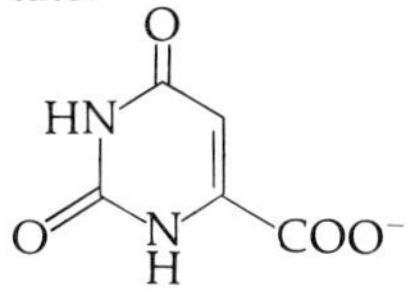

Orotate

[C:341]

427 Reaction with PRPP transforms orotate into a nucleotide that undergoes decarboxylation to give UMP.
[C:341]

428 Mitochondrial carbamoyl phosphate is fed into the urea cycle, whilst cytoplasmic carbamoyl phosphate is required for pyrimidine biosynthesis. This probably ensures entirely separate control of amino acid degradation and pyrimidine nucleotide synthesis.
[F:519]

429 Kinases exist in cells that enable ATP to phosphorylate other nucleoside monophosphates to give di- and triphosphates.
[F:521]

430 The amination of UTP to give CTP is effected by glutamine. The CTP synthetase enzyme also requires ATP for this UTP → CTP conversion to occur.
[F:521]

## C Deoxyribonucleotides

431 The 2'-deoxyribonucleotides are formed by reduction of the nucleoside diphosphates. This is a complex process that, in humans, involves the simultaneous oxidation of a reduced form of the cofactor thioredoxin.
[F:524]

432 A formyl derivative of $FH_4$ (actually $N^5$, $N^{10}$-methylene $FH_4$) acts to convert the 2'-deoxy derivative of UMP (i.e., dUMP) into TMP in the presence of thymidylate synthetase. The process involves the insertion of a methyl group into position 5 of the pyrimidine ring. A phosphatase acts on dUDP to give the dUMP.
[F:526]

433 The nucleotides can be synthesized *de novo* in several tissues, but the main site is the liver. Most cells can employ salvage pathways to synthesize required nucleotides.
[G:762, 768]

434 Not only must control be exercised on the overall rate of nucleotide biosynthesis, but proper balance between synthesis of the individual nucleotides is essential. To cover these requirements, not only can interconversion of nucleotides occur, but a complex system of allosteric regulation of enzyme activity and of repression and depression of enzyme synthesis provides delicate control.
[C:339, 344]

435 Cytosine arabinoside and azacytidine are used to treat leukaemias and 5-fluorouracil and 5-fluorodeoxyuridine monophosphate are used to treat carcinomas. 5-Iododeoxyuridine is an antiviral agent used to treat herpetic and vaccinial disease.
[A:536]

## D Uric acid, gout and the Lesch–Nyhan syndrome

436 Nucleases effect the hydrolysis of nucleic acids and phosphatases convert the resulting nucleotides into nucleosides. The latter are further broken down by the nucleosidases or nucleoside phosphorylases to give purine or pyrimidine bases.
[C:331, 385]

437 Purine nucleotides undergo various phosphorolytic and deaminative processes to give hypoxanthine or xanthine. The hypoxanthine is oxidized to xanthine by xanthine oxidase. Further action of xanthine oxidase converts xanthine into uric acid. For example,

Guanine (4-amino-hypoxanthine)

Xanthine                    Uric acid

[C:332]

438 Xanthine oxidase is a flavoprotein.
[A:514]

439 The drug allopurinol is a close chemical analogue of hypoxanthine (groups at positions 7 and 8 of hypoxanthine are transposed). Xanthine oxidase effects the oxidation of allopurinol to give oxypurinol, the analogue of xanthine with groups at positions 7 and 8 transposed.
[A:520]

440 Abnormally high concentrations of urate ions in blood and of uric acid (or sodium urate) in urine are characteristic of gout. The condition may be symptomless. It may, however, give rise to painful arthritic attacks and to the formation of uric acid kidney stones. The disease is more common in males over 40 years old.
[A:517]

441 Small crystals of sodium urate are engulfed by polymorphonuclear leucocytes, but the urate structure forms hydrogen bonds with leucocyte membrane components. This causes membrane damage and the release of lysosomal enzymes that attack surrounding tissue.
[A:518]

442 Metabolic gout arises from an overproduction of uric acid and its precursors. This may be caused by an inherited defect in an enzyme (such as PRPP synthetase) or it may be a secondary consequence of some other disease. Renal gout arises from underexcretion. This may be the result of a defect in an enzyme concerned with uric acid (or urate) transport.
[A:518]

443 Colchicine and sometimes other anti-inflammatory drugs are helpful to gouty patients. Allopurinol inhibits the xanthine oxidase-catalysed oxidation of uric acid precursors and thus reduces uric acid production.
[A:519]

444 The Lesch–Nyhan syndrome is an inherited X-linked recessive disease affecting males. There is greatly increased uric acid formation and excretion, leading to gout symptoms. In addition, retarded mental development, spasticity and choreoathetosis are evident and a strong tendency to commit self-mutilation is manifest.
[C:347]

445 Hypoxanthine-guanine phosphoribosyl transferase acts to salvage a proportion of purines that would otherwise be oxidized to uric acid. Catalysed by this enzyme, PRPP converts these purines into nucleotides. The virtual absence of this enzyme activity in the Lesch–Nyhan syndrome leads to excessive uric acid formation and to gout symptoms. The biochemical basis of the neurological manifestations is unknown.
[A:525]

# NUCLEIC ACIDS AND PROTEIN SYNTHESIS

## A DNA structure and replication

446 Each DNA chain is a polydeoxyribonucleotide made up of $5'$-phosphonucleoside residues, joined from the $3'$ to the $5'$ positions by phosphodiester bridges. The nucleoside portions of these residues are selected from adenosine, guanosine, thymidine and cytosine.
[F:559]

447 The DNA structure contains 2 antiparallel chains (one $5' \rightarrow 3'$ and the other $3' \rightarrow 5'$) joined by hydrogen bonds between complementary purine and pyrimidine bases (A===T and G≡≡≡C) present in the complementary chains. The structure is arranged as a right-handed double helix, with the base pairs crossing the common axis of the double helix and with the deoxyribose phosphate groups on the outside. The main type of helical structure in solution is described as a B-helix. The exterior is studded with negative charges residing on the phosphate groups.
[F:565]

448 Deoxyribonucleoside triphosphates polymerize in the presence of DNA polymerase, $Mg^{2+}$ ions, a primer and a DNA template. The four triphosphates, dATP, dTTP, dGTP and dCTP must be present. DNA polymerase I adds each new nucleotide to a 3'-hydroxy terminus of an *existing* DNA or RNA strand and, consequently, a primer chain with a free 3'-OH group is required. The DNA template itself may be a long single strand of DNA or a double-stranded DNA with a single-stranded gap. The (short) primer strand must associate by base pairing with the template single chain.
[A:385]

449 A DNA ligase enzyme can repair a break in a DNA chain that is part of a double-stranded DNA, i.e. it can suitably activate, and then link, a 5'-phosphate group to an adjacent, but separate, 3'-hydroxy group. The nucleotides to be linked must be paired to adjacent bases on a complementary strand. The presence of ATP or another activating molecule is required.
[A:389]

450 Double-stranded DNA in some bacteria (including *E. coli*), and in some viruses and mitochondria, consists of a closed loop described as 'circular' DNA.
[F:573]

451 Evidence indicates that DNA chain synthesis occurs in the $5' \rightarrow 3'$ direction only.
[A:386]

452 At first, a portion of the very long double helical

DNA is unwound and the complementary strands separated by the action of specific protein molecules. The 'bubble' that forms in the double helical structure grows in size in advance of a moving 'replicating fork', where synthesis is started at specific loci by a DNA-dependent RNA polymerase. At each locus, this enzyme generates a short complementary chain of RNA primer and then, catalysed by DNA polymerase, deoxyribonucleotides add to the 3' end of this primer to produce a segment of new DNA chain (an Okazaki fragment). Synthesis proceeds in short bursts *in opposite directions* along the 2 separating strands at the moving replicative fork. Then, RNA primer segments are lopped off, the Okazaki fragments extend to fill the short gaps, and finally DNA ligase acts to join the (1000 or 2000 residue) DNA fragments. Thus, a new complementary strand is synthesized for each of the 2 original strands in the DNA helix.
[F:583]

453 All 3 DNA polymerases, under suitable conditions, can (a) catalyse template directed DNA synthesis in the $5' \rightarrow 3'$ direction and (b) remove by hydrolysis, a DNA 3' terminal nucleotide (i.e., possess $3' \rightarrow 5'$ exonuclease activity). Polymerases II and III differ from I in (a) lacking $5' \rightarrow 3'$ nuclease activity and (b) preferring relatively short, separated lengths of single-stranded DNA as templates.
[F:580]

454 The sequence order of the 4 nucleotides A, C, G, T in the DNA chain carries the genetic information. A group of 3 bases, called a codon, codes for each of the 20 amino acids.
[F:620]

455 If segments of DNA from one organism are spliced onto extrachromosomal pieces of DNA from another organism, a recombinant DNA is produced. This may be able to replicate rapidly in cells of the second organism and provide a method for the synthesis of a particular DNA in relatively large quantities.
[F:760]

## B  Biosynthesis of RNA

456 The three types, messenger RNA (mRNA),

ribosomal RNA (rRNA) and transfer RNA (tRNA) are found in all cells. Eucaryotic cells also contain nuclear RNA (nRNA), a type that is less well characterized.
[F:599]

457 An RNA chain is similar to a DNA strand in being a long unbranched polynucleotide with its residues linked by $3' \rightarrow 5'$ phosphodiester bonds. In mRNA three of the nucleotides (A, C and G) are identical with those in DNA, but the fourth is U, not T. Other bases are present in tRNA structures. The pentose in RNA is ribose, not deoxyribose as in DNA. Finally, except for some viral RNAs, RNA structures usually occur as single chains rather than double-stranded forms.
[F:598]

458 Synthesis of mRNA occurs at a DNA template. All four triphosphates ATP, UTP, GTP and CTP must be present together with $Mg^{2+}$ (which may be partly replaced by $Mn^{2+}$) and DNA-dependent RNA polymerase.
[F:603]

459 RNA synthesis proceeds in the $5' \rightarrow 3'$ direction as with DNA synthesis.
[F:609]

460 RNA polymerase is found in the nucleus, nucleolus and in mitochondria.
[A:407]

461 The reaction

$n$ ribonucleotide triphosphates

$$\Updownarrow$$

RNA + $n$ pyrophosphate ions

is reversible, but the presence of a pyrophosphatase keeps the pyrophosphate level low. This minimizes the reverse reaction.
[A:407]

462 The detailed mechanism of transcription reactions is not fully understood. It is probable that only one of the 2 DNA strands acts as a template, although the template function could switch from one strand to the other. The RNA synthesized has a structure complementary to that of

the template strand. Single-stranded DNA is a very inefficient template.
[F:608]

463   Some RNA viruses have been shown to possess an RNA-dependent DNA polymerase that synthesizes DNA on an RNA template (reverse transcriptase). The enzyme uses viral RNA to form a DNA–RNA hybrid. The RNA strand is removed, and replication of the DNA gives a double-stranded DNA that acts as a template for the replication of viral RNA.
[F:743]

## C   Activation of amino acids

464   Either G or C is present at the 5' end of the tRNA chain, whilst all tRNA molecules have –C–C–A as the terminal sequence at the 3' end.
[A:404]

465   The tRNA structure is a single chain with less than 100 nucleotide residues, many of which are 'unusual' in the base portion, e.g., bases such as hypoxanthine, 5-methylcytosine and isopentyl-adenine may be present. A hydrogen-bonded clover leaf structure is adopted, with chain ends on one 'leaflet' and an anticodon of 3 specific nucleotides at the tip of the opposite 'leaflet'.
[F:647]

466   Both tRNA molecule and aminoacyl-tRNA synthetase are specific for each particular amino acid. Reaction occurs at the free 3'-hydroxy group of the terminal –C–C–A sequence as follows:

$$(tRNA)^\alpha-C-C-A-OH + HO-CO-CHR^\alpha NH_2$$

$$\downarrow \quad \begin{array}{l} \nearrow ATP \\ \searrow AMP + PP_i \end{array}$$

$$(tRNA)^\alpha-C-C-A-O-CO-CHR^\alpha NH_2$$

[F:643, 649]

467   The tRNA molecule must be able to recognize (a) the particular aminoacyl-tRNA synthetase, (b) a binding site on a large ribosomal subunit, and (c) the correct three nucleotide codon on a mRNA molecule.
[A:406]

468    Although each amino acid has at least one cor-
       responding tRNA, some amino acids appear to
       have available to them more than one tRNA in
       some cells.
       [A:402]

## D    Coding and translation

469    The code consists of a triplet of adjacent non-
       overlapping ribonucleotide bases for each par-
       ticular amino acid.
       [F:621]

470    As there are only 4 bases (A C G U) the number
       of possible arrangements in pairs is $4^2$ or 16; not
       enough to code for 20 amino acids.
       [F:620]

471    The triplet code provides $4^3$ or 64 possible
       'words'. Now, either most of these are nonsense
       words or the code is degenerate in that some, if
       not all, amino acids have more than one codon.
       Only methionine has a single codon (AUG)
       whilst at least 2 codons code for each of the other
       19 amino acids.
       [F:622]

472    The 3 triplets (UAA, UAG and UGA) are chain-
       terminating triplets that signal the completion of
       a newly synthesized polypeptide chain.
       [F:628, 661]

473    The codon (AUG) that codes for both N-formyl-
       methionine (fMet) and for methionine (Met) is
       also the chain-initiating codon. fMet is concerned
       with chain initiation in E. coli and other bacteria,
       whilst Met is its counterpart in eucaryotic cells.
       When synthesis is complete, the formyl group
       and in some cases the methionine residue are
       cleaved from the polypeptide N-terminus. One of
       the valine codons (GUG) may sometimes be
       involved in the initiation of polypeptide chain
       synthesis.
       [F:630, 656]

474    An inactive 70S ribosome dissociates into 2 sub-
       units and then mRNA, fMet-tRNA and GTP
       associate with the smaller 30S ribosomal sub-
       unit. Three protein initiation factors are also
       involved in the formation of this 30S complex.

The larger (50S) ribosomal subunit then associates with the smaller unit to form a 70S initiation complex ready for protein synthesis. The protein initiation factors and GDP + inorganic phosphate are released. The fMet-tRNA is located so that its anticodon triplet pairs (by specific hydrogen bonding) with an initiating codon AUG (or GUG) on the mRNA, which is attached to the smaller subunit. The fMet-tRNA is also associated with a site (P site) situated on the larger subunit.
[A:428]

475 A second specific aminoacyl-tRNA possesses an anticodon that recognizes the codon triplet adjacent to the initiating codon. This tRNA is also located by a second site (A site) on the large ribosome. A protein elongation factor and GTP are required for this association to occur. The new amino acid residue is adjacent to the fMet residue and a peptidyl transferase, present in the 50S subunit, transfers the fMet residue onto the second amino acid residue forming the new peptide bond. No ATP or GTP is required for this step.
[A:432]

476 After the first peptide bond is formed, the P (peptidyl) site is quickly vacated by the tRNA and the adjacent A (aminoacyl) site is temporarily occupied by the newly formed peptidyl-tRNA. Translocation moves the latter, together with its attached mRNA chain, in relation to the ribosome subunits, so that the peptidyl-tRNA occupies the P site and vacates the A site. The latter is then ready for the next specific aminoacyl-tRNA. Chain elongation then proceeds in this manner. Energy for each translocation is provided by a $GTP \rightarrow GDP + P_i$ reaction.
[A:434]

477 No aminoacyl-tRNA will recognize a terminating codon, but these are recognized by protein release factors that then block site A and activate peptidyl transferase, which acts to release the completed polypeptide.
[A:435]

478 In transcription, the RNA polymerase reads the

sequence of a DNA strand in the 3' to 5' direction, but the mRNA chain formed grows in the 5' to 3' direction. In protein synthesis, the mRNA is read from the 5' to 3' direction, producing a polypeptide chain that grows in the sense indicated by $H_3\overset{+}{N} \dashrightarrow COO^-$.
[A:609, 655]

479 A mutation is a change in the base sequence of DNA. In a transition mutation (a copy error) one purine–pyrimidine base pair is replaced by another (e.g., A===T replaced by G≡≡≡C), whilst in a transversion mutation one purine-pyrimidine pair is replaced by a pyrimidine-purine (e.g., A===T replaced by T===A). These mutations affect one codon. A frame-shift mutation arises from the insertion or deletion of one or more nucleotides. Such mutations affect all information in the strand from the point of mutation to the next initiation or control site.
[F:635]

480 Polysomes are clusters of ribosomes connected by a strand of mRNA. This complex permits simultaneous translations from one mRNA chain.
[F:655]

481 Post-ribosomal modifications of proteins include (a) the removal of a formyl group and of one or more N-terminal amino acids; (b) other types of cleavage of a long polypeptide chain; (c) the formation of disulphide groups; (d) the modification of side chain R groups by hydroxylation, phosphorylation, methylation and the attachment of other ligands; (e) the attachment of one or more prosthetic groups to a protein.
[F:661]

482 Secretory protein is made by ribosomes that are bound in rows to the outer surface of parts of the endoplasmic reticulum (ER). The large subunit is attached to the ER and the growing peptide chains extend through into the cisternae of the reticulum. Completed chains are then released into the cisternae where they may be transported for further processing and subsequent secretion.
[F:712]

483 The classic case is control in *E. coli* of the synthesis of enzymes required for the metabolism of lactose. The DNA operon concerned with synthesis of these proteins consists of a promoter segment that initiates synthesis of the appropriate mRNA, an operator segment, and the appropriate structural genes that code for the enzymes. The operon is normally switched off because of the attachment of a specific regulator protein (repressor) to the operator segment. The repressor is coded for by a separate regulator gene. If lactose is present, a small proportion will be converted into allolactose, which acts as an inducer. Its molecules bind to repressor molecules, removing their affinity for the operator segment and, consequently, their power to block synthesis of those enzymes required for the degradation of lactose.
[F:669]

## E Mechanism of action of some antibiotics

484 Actinomycin binds tightly to the DNA double helix where guanine is present. Its molecules are inserted (intercalated) between adjacent sets of base pairs in the double helix, the effect being to strongly inhibit RNA synthesis. Synthesis of mRNA is mainly affected because of its rapid rate of turnover. Replication of DNA is not inhibited at low actinomycin concentrations.
[F:613]

485 Actinomycin D is an antineoplastic agent used in the treatment of Wilm's tumour. In addition, its ability to block transcription has proved of great value in research. Thus, if the replication of an RNA virus is being studied, the host cells may be treated with actinomycin D before being infected. The cells will then synthesize viral protein almost exclusively.
[A:409, 502]

486 Antibacterial and antiviral activity is exhibited by rifamycin (from *Streptomyces*) and by some chemically modified variants, one of which, rifampicin, is used therapeutically. The double-stranded DNA template is not damaged, but interference occurs with the formation of the phosphodiester bond that would link the first 2 residues of a new RNA chain. Elongation of

existing (short or long) chains is not prevented.
[A:409]

487 Puromycin is the $p$-methoxyphenylalanyl derivative of a chemical analogue of adenosine and is
analogous to the terminating aminoacyl-adenosyl
residue of an aminoacyl-tRNA. It enters the
ribosomal site A and its primary amino group
froms a peptide bond with the newly synthesized
polypeptide. However, subsequent processes
cannot occur and polypeptide chain synthesis
ceases.
[F:663]

488 Streptomycin binds to the 30S ribosomal subunit
and probably alters its conformation. Interference with the binding of formylmethionyl-tRNA
inhibits the initiation of polypeptide chain synthesis and weakened anticodon–codon bonding
produces misreading of the code.
[A:440]

489 These 2 antibiotics inhibit the peptidyl transferase activity of particular ribosomal subunits. The
activity of 50S subunits (in prokaryotes and in
the mitochondria of eukaryotes) is inhibited by
chloramphenicol and that of the 60S subunits (in
eukaryotes) is inhibited by cycloheximide.
[A:440]

490 Other inhibitors of protein synthesis include
neomycin, erythromycin and tetracyclines.
[A:440]

# LIPIDS AND STEROIDS

## A  Membrane lipids

491 The major lipid classes in membranes are phospholipids, glycolipids and cholesterol.
[F:207]

492 Glycerophospholipids have the general structure

$$\begin{aligned}&CH_2O\text{--}R'\\&CHO\text{--}R\\&CH_2O\text{-}PO_2^-\text{--}X\end{aligned}$$
where R and, except in plasma-
logens, R' are fatty acid residues
$-CO-(CH_2)_n-CH_3$

153

Subgroups are:
(a) lecithins (phosphatidyl cholines), $X$ is the group $-OCH_2CH_2N(CH_3)_3$,
(b) cephalins, $X$ may be $-OCH_2CH_2NH_2$, $-OCH_2CH(NH_2)COOH$ or an inositol residue,
(c) plasmalogens, $R = -OCH=CH-(CH_2)_n-CH_3$ (forming an ether linkage), $X$ is a choline, ethanolamine or serine residue,
(d) polyglycerol phospholipids.
[A:844, 849]

493 Sphingolipids have the structure:

$$CH_3-(CH_2)_{12}-CH=CH-CHOH$$
$$\mid$$
$$CH.NH-R$$
$$\mid$$
$$CH_2O-Y$$

where $R = -CO-(CH_2)_n-CH_3$.
Subgroups are:
(a) ceramides, $Y = H$,
(b) cerebrosides, $Y = $ a hexose residue, e.g., a $1\text{-}\beta\text{-D-galactoside}$ group,
(c) sphingomyelins, $Y = -PO_2^- - X$, where $X$ is a choline or ethanolamine residue.
[F:461]

494 Phosphatidic acids are converted by CTP into active CDP-diglycerides, which react readily with alcohols such as serine or inositol giving phosphatidyl serines or inositols respectively and releasing CMP. Phosphatidyl serines are decarboxylated to give the ethanolamines, which give the cholines on methylation. In sphingolipid synthesis, palmitoyl CoA is reduced to palmitaldehyde, which condenses with serine to give dihydrosphingosine. Dehydrogenation with FAD produces sphingosine. The $N$-acyl group is introduced by reaction with fatty acyl CoA to give a cerimide. Reaction with CDP-choline gives sphingomyelin and with UDP-hexose gives cerebroside. Thus CTP is used to form CDP-activated hydroxy compounds and fatty acyl CoA derivatives are the activated fatty acids. Serine forms part of the structure of sphingosine and of several glycerophospholipids.
[F:458]

495 Membrane phospholipid and glycolipid is arranged in a bilayer form with lipophilic chains of both layers oriented in an approximately parallel fashion within the membrane and with the hydrophilic groups on the two external faces. The arrangement is not symmetrical across the membrane; thus, in erythrocytes, phosphatidyl cholines are on the external surface, whilst the ethanolamines and serines are on the cytoplasmic surface. Protein molecules are imbedded in the membrane, some piercing both faces. There is lateral (2-dimensional) movement of membrane components, but very restricted 'vertical' movement with no rotations in 'vertical' planes.
[F:224]

496 Cholesterol is a major component of the uncharged lipid present in the plasma membranes of animal cells. Presence of cholesterol molecules within the membrane can disorganize the well organized, tightly packed array of parallel chains that exists with saturated chains and render such a region more fluid in character. The presence of *cis* double bonds in chains produces changes in direction and such chains will be less organized and the membrane region more fluid. Presence of cholesterol molecules may, however, fill molecular gaps and help to gel such a region.
[F:226]

497 Carbohydrate chains attached to the cell membrane outer surface have considerable specificity and are concerned in the recognition of molecules or bodies that arrive at the external surface, and in antigenic and receptor roles towards such entities.
[F:221]

498 Lipid-soluble substances can usually 'dissolve' readily in the membrane lipid and cross the membrane by passive diffusion. However, many polar substances can also cross the membrane leaflet by the agency of specific permeases that can selectively operate channels across the membrane. Certain polar substances can then negotiate passage through these channels. In both purely passive and in facilitated diffusion (above), transport occurs only down a concentration gradient. Active diffusion of ions, and of

many other polar substances, occurs against concentration gradients. An appropriate enzymic mechanism, utilizing ATP or another high energy compound is necessary for such active transport.
[F:861]

499 The following diseases are examples of lipidoses: Niemann–Pick, Tay–Sachs, Gaucher's, Krabbe's and Fabry's disease and generalized gangliosidosis. They are generally characterized by the accumulation of particular lipids in tissues, often in brain, liver, spleen and bone marrow, and by mental retardation and neurological disease. The defect in Niemann–Pick disease is of sphingomyelinase, leading to the accumulation of sphingomyelins in brain, liver and spleen. Mental retardation occurs and prognosis is very poor.
[A:861]

500 Respiratory distress syndrome (RDS) is one of the principal causes of death in premature infants and is sometimes fatal in full-term infants. An increase in lecithin concentration in fetal lung parallels the development of resistance to RDS that occurs with increasing lung maturity. At week 35 of normal development, a rapid rise in lung lecithin synthesis commences and the risk of RDS rapidly decreases.
[A:847]

# B  Cholesterol

501 The cholesterol molecule possesses the steroid fused ring system, consisting of 3 6-membered rings ($A$, $B$ and $C$) and a 5-membered ring $D$. A double bond is present in ring $B$ (between $C_5$ and $C_6$) and 4 substituents are present as follows:
(a) a hydroxy group in ring $A$ (at position 3),
(b) an angular methyl group between rings $A$ and $B$ (at position 10),
(c) an angular methyl group between rings $C$ and $D$ (at position 13),
(d) a 1,5-dimethylhexyl group in ring $D$ (at position 17).
All 4 substituents are located on the same side of the steroid ring system and thus all 4 have the same designation ($\beta$). The chemical structure of

cholesterol is given for reference:

[C:192]

502 Cholesterol is present in all mammalian cells, especially in nerve tissue. It is a component of animal, but not of plant, lipid.
[C:192]

503 Cholesteryl esters may form in (a) liver, where reaction may occur between a fatty acyl CoA and cholesterol; (b) plasma, where lecithin-cholesterol acyl transferase (LCAT) can catalyse the transference of an *unsaturated* fatty acyl group from lecithin to cholesterol; (c) intestinal mucosa, by reaction between cholesterol and fatty acid anions.
[D:453]

504 The main site of cholesterol biosynthesis is the liver, with smaller contributions from the intestine, adrenal cortex and other tissues. The unit from which the cholesterol molecule is constructed is the acetyl group (of acetyl CoA).
[D:454]

505 Two acetyl CoA molecules condense, with the elimination of CoA, to form acetoacetyl CoA. Then a further acetyl CoA molecule reacts with the elimination of CoA to give a molecule of β-hydroxy-β-methylglutaryl CoA (HMG CoA). This compound is converted into ketone bodies in mitochondria, but in the cytosol it is used for cholesterol synthesis.
[C:237]

506 The key step in the biosynthesis of squalene is the conversion of HMG CoA to mevalonic acid,

as shown:

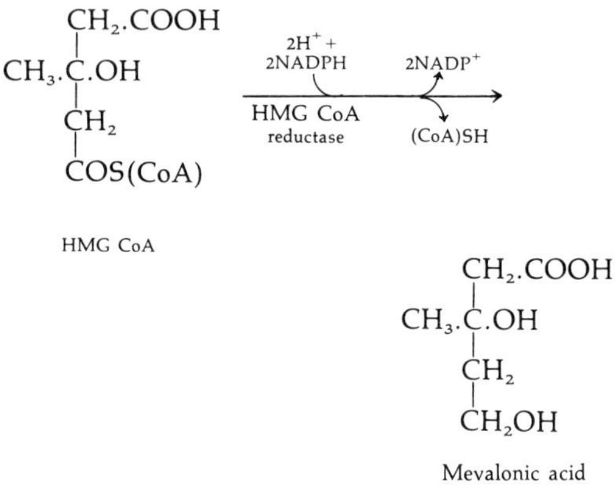

$$CH_3\text{-}\underset{\overset{|}{COS(CoA)}}{\underset{|}{\overset{|}{\underset{|}{C}}}}\text{.OH}$$

CH₂.COOH
|
CH₃.C.OH
|
CH₂
|
COS(CoA)

$\xrightarrow[\substack{\text{HMG CoA} \\ \text{reductase}}]{\substack{2H^+ + \\ 2NADPH \qquad 2NADP^+}}$ (CoA)SH

HMG CoA

CH₂.COOH
|
CH₃.C.OH
|
CH₂
|
CH₂OH

Mevalonic acid

[C:237, 240]

507 Isoprenoid units are the 5-carbon units that are the building blocks for the construction of the steroid molecule. They exist as the 2 isoprenoid isomers:

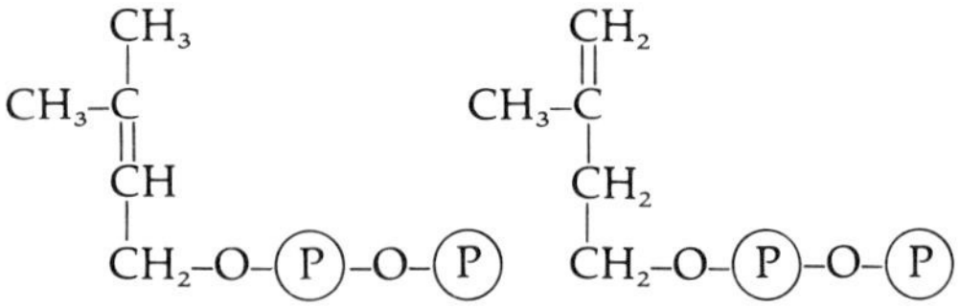

CH₃
|
CH₃–C
‖
CH
|
CH₂–O–(P)–O–(P)

CH₂
‖
CH₃–C
|
CH₂
|
CH₂–O–(P)–O–(P)

An isomerase effects the interconversion of the 2 isomers. They are formed from mevalonic acid by successive phosphorylations with ATP to give mevalonic acid 5-pyrophosphate and then an unstable phosphorylated intermediate that decomposes to give isoprenoid units as follows:

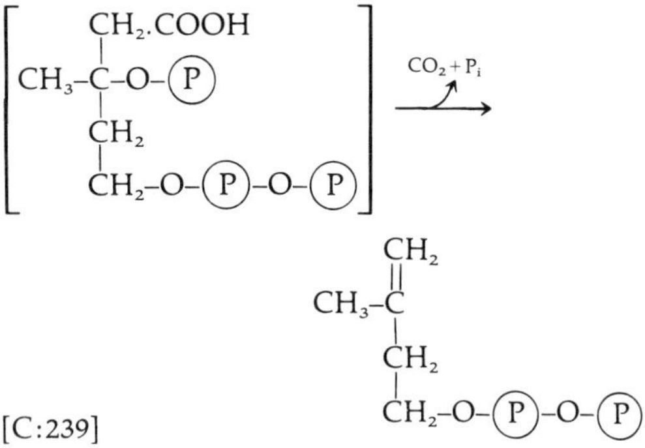

⎡ CH₂.COOH ⎤
⎢ | ⎥
⎢ CH₃–C–O–(P) ⎥
⎢ | ⎥
⎢ CH₂ ⎥
⎣ CH₂–O–(P)–O–(P) ⎦

$\xrightarrow{CO_2 + P_i}$

CH₂
‖
CH₃–C
|
CH₂
|
CH₂–O–(P)–O–(P)

[C:239]

508 The 2 isoprenoid isomers can condense together with elimination of pyrophosphate to form the 10C unit geranyl pyrophosphate. The skeleton then builds up as follows:

Geranyl pyrophosphate (10C)

    ↓ + Isoprenoid isomer (5C)

Farnesyl pyrophosphate (15C)

    ↓ + Farnesyl pyrophosphate (15C)

Squalene (30C)

Squalene (30C) is an open chain compound that is converted into lanosterol (30C) by ring closures. This is an oxidative process requiring molecular oxygen and NADPH. The conversion of squalene to lanosterol is probably a second control step in cholesterol biosynthesis. The synthesis then proceeds by two routes, *via* 7-dehydrocholesterol or *via* desmosterol. Both routes involve the removal of three methyl groups to give the correct 27C carbon skeleton of cholesterol and then various reactions to obtain the correct single unsaturation in ring *B*.
[C:239]

509 The main regulatory enzyme in cholesterol biosynthesis is HMG CoA reductase. The production of this enzyme in liver is inhibited by high dietary cholesterol (probably acting through a cholesterol-containing plasma lipoprotein). Its activity is also inhibited by some cholesterol metabolites including $7\alpha$-hydroxycholesterol and $20\alpha$-hydroxycholesterol, which are intermediates in the biosynthesis of bile salts and steroid hormones respectively. Various steroid and peptide hormones also exert regulatory effects on the HMG CoA reductase activity of liver cells.

A diet, high in saturated fat, appears to stimulate cholesterol biosynthesis. However, the replacement of saturated fatty acids in dietary fats by polyunsaturated fatty acids produces a hypocholesterolaemic effect.
[C:240]

510 Atherosclerosis is a degenerative condition of arterial walls in which cholesteryl esters and

other lipids are deposited in the connective tissue. Occlusion of blood vessels supplying oxygen to heart muscle leads to coronary attacks and occlusion of blood vessels supplying oxygen to brain leads to strokes. There is a correlation between total serum cholesterol (with its fatty acyl esters) and atherosclerosis. However, the apparently damaging effect of high serum cholesterol is dependent on the type of lipoprotein with which it is associated. Cholesterol associated with high density lipoprotein (HDL) appears to have an inverse correlation with coronary disease.
[C:241]

## C  Bile acids

511  There is some lack of consistency in the use of the term bile acid, but it is generally taken to cover:
(a) the primary bile acids (cholic acid and chenodeoxycholic acid);
(b) the conjugated bile acids (glycine, taurine and other conjugates of the primary bile acids);
(c) the secondary bile acids (deoxycholic and lithocholic acid);
(d) conjugates of the secondary bile acids.
[A:923]

512  Cholic acid and chenodeoxycholic acid are synthesized as their CoA derivatives from cholesterol in liver.
[A:923]

513  Cholic acid is a $C_{24}$ steroid containing an OH group in each of rings $A$, $B$ and $C$. Chenodeoxycholic acid is similar, but lacks the OH group that is present in ring $C$ (at position 12) of cholic acid. The first, and rate-controlling, step of the synthesis is the introduction of a hydroxy group into ring $B$ (at position 7) of cholesterol. 7-Hydroxycholesterol then undergoes a reaction sequence that removes the double bond in ring $B$, gives the required stereochemical configurations, and effects partial hydroxylation at position 12 (in ring $C$). The final processes, shown below, involve the hydrocarbon side chain on ring $D$. Oxidation removes the terminal isopropyl group

and gives the CoA derivative of either cholic or chenodeoxycholic acid. The side chain modification is as follows:

$$R-CH(CH_3).CH_2.CH_2.CH_2.CH(CH_3)_2$$

$$\downarrow$$

$$R-CH(CH_3).CH_2.CH_2.COS(CoA)$$

[A:924]

514 Before secretion in bile, the CoA derivatives of cholic and chenodeoxycholic acids react with glycine and taurine. Thus cholyl CoA gives glycocholic and taurocholic acids:

Cholyl–S(CoA)

$+H_3\overset{+}{N}CH_2COO^-$    $(CoA)SH + H^+$    $+H_3\overset{+}{N}CH_2CH_2SO_3^-$

Cholyl–NHCH$_2$COO$^-$    Cholyl–NHCH$_2$CH$_2$SO$_3^-$

These conjugates are stronger acids than the primary bile acids and they are considerably more soluble in water. They will be present as anions in the slightly alkaline bile.
[A:926]

515 Deconjugation of bile salts occurs intestinally and some of the primary bile acids released get reduced by the action of bacterial enzymes. This reduction removes the 7-hydroxy group, converting cholic acid into deoxycholic acid and chenodeoxycholic acid into lithocholic acid. Deoxycholic acid and lithocholic acid are often called secondary bile acids.
[A:927]

516 Two main controls on the biosynthesis of bile acids in liver arise in the following way:
(a) Over 90% of bile acids, secreted in bile, are reabsorbed, mainly from the ileum by active transport, but also from the jejunum and colon. The reabsorbed bile acids are returned to the liver by the portal circulation. Biosynthesis of new bile acid decreases as bile acid return increases.
(b) If the intestinal absorption of cholesterol increases, transport of cholesterol to the liver

161

increases. This cholesterol stimulates the synthesis of bile acids.
[A:923, 925]

517 Small quantities of cholesterol are metabolized to give steroid hormones and cholecalciferol. Cholesterol may also be esterified and it may be reduced to cholestanol. However, the steroid ring system cannot be catabolysed to give $CO_2$ and water and consequently the elimination of cholesterol depends largely on the conversion of cholesterol into bile acids. About 300 mg of bile acids are eliminated daily in the faeces and this probably facilitates the faecal elimination of a greater quantity of cholesterol itself and of derived neutral steroids.
[A:919]

518 Both primary and secondary bile acids are reabsorbed intestinally and transferred to the liver. Cholesterol metabolism only produces CoA derivatives of primary bile acids, but because both primary and secondary bile acids reach the liver, both may be converted into CoA derivatives, undergo conjugation, and be secreted as conjugates into bile.
[I:299]

## D Steroid hormones

519 The main groups of steroids with their main sites of production are as follows:
(1) progestogens (corpus luteum)
(2) androgens (testis)
(3) oestrogens (ovarian follicle)
(4) glucocorticoids (adrenal cortex)
(5) mineralocorticoids (adrenal cortex)
[F:474]

520 An example of each type of steroid hormone with an indication of its function follows:
(1) *progesterone*, prepares the lining of the uterus for implantation of an ovum and provides conditions necessary for the continuation of a pregnancy;
(2) *testosterone*, acts to develop male secondary sex characteristics;
(3) *oestradiol*, acts to develop female secondary

162

sex characteristics and plays an important role in the ovarian cycle;

(4) *cortisol*, induces gluconeogenesis and glycogenesis, stimulates the catabolism of muscle, epidermal and connective tissue, stimulates the release of glycerol and fatty acid from adipose tissue, and exerts anti-inflammatory and antiallergic activities;

(5) *aldosterone*, stimulates renal reabsorption of $Na^+$ (together with $Cl^-$ and $HCO_3^-$), a corollary of which is increase in blood volume and blood pressure.

[B:726]

521 The 1,5-dimethylhexyl side chain, attached to ring *D* of cholesterol, is oxidized to an acetyl side chain in pregnenolone ($C_8$ side chain $\rightarrow$ $C_2$ side chain). This process occurs by 2 consecutive hydroxylations in the side chain, followed by oxidative cleavage. All 3 reactions require molecular oxygen and NADPH. The process can be summarized as:

$$R–CH(CH_3)CH_2CH_2CH_2CH(CH_3)_2$$

$$\downarrow$$

$$\underset{\underset{OH}{|}}{R–C}(CH_3).\underset{\underset{OH}{|}}{CH}.CH_2CH_2CH(CH_3)_2$$

$$\downarrow$$

$$R–COCH_3 + OCH.CH_2CH_2CH(CH_3)_2$$

[F:475]

522 Adrenocorticotropic hormone (ACTH) stimulates the conversion of cholesterol into pregnenolone. ACTH is a polypeptide originating from the anterior pituitary gland.
[F:476]

523 In the formation of progesterone, the 3-hydroxy group of ring *A* of pregnenolone is oxidized to a ketone and the double bond rearranges from the 5,6 position of ring *B* to the 3,4 position of ring *A*, so that it forms a conjugated double bond system C=CH–C=O with the ketone group in

163

ring *A*. The transformation is as follows:

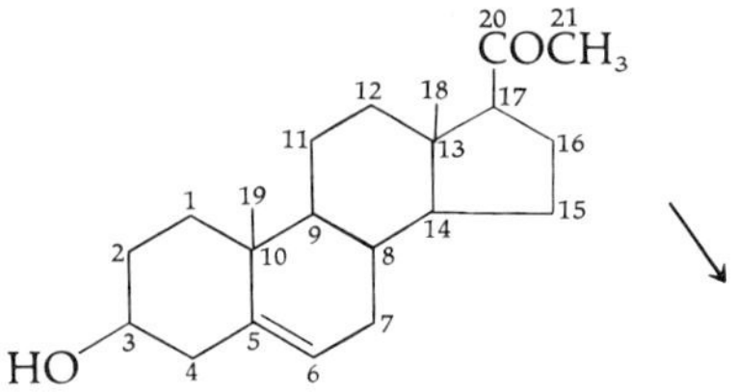

Pregnenolone

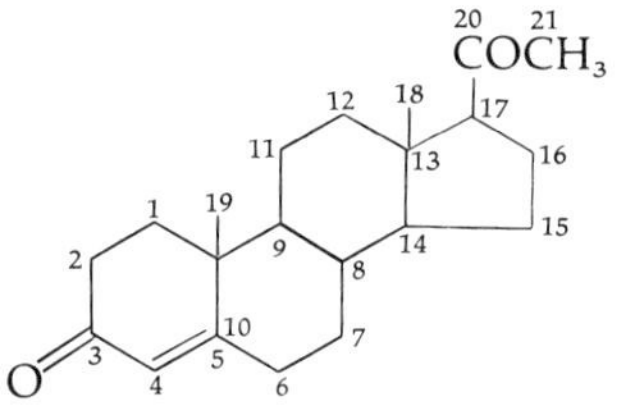

Progesterone

[F:476]

524 Hydroxylations occur in the progesterone structure to give 11,21-dihydroxyprogesterone (corticosterone) and 11,17,21-trihydroxyprogesterone
(cortisol). Cortisol is the major corticosteroid.
[F:476]

525 The angular methyl group, containg C atom 18,
situated between rings *C* and *D* of corticosterone
is oxidized to give a formyl group (–CHO) in
aldosterone.
[F:476]

526 Progesterone is converted into other sex hormones by the following processes. Hydroxylation
at position 17 is followed by cleavage of the $C_{17}$
side chain to give the androgen, androstenedione.
Further sex hormones are obtained from androstenedione. Thus, reduction of the 17-oxo group
to a 17-hydroxy group gives testosterone.

Oxidation of androstenedione removes an angular methyl group and introduces a further
double bond to convert ring *A* into an aromatic

ring. The product formed is oestrone:

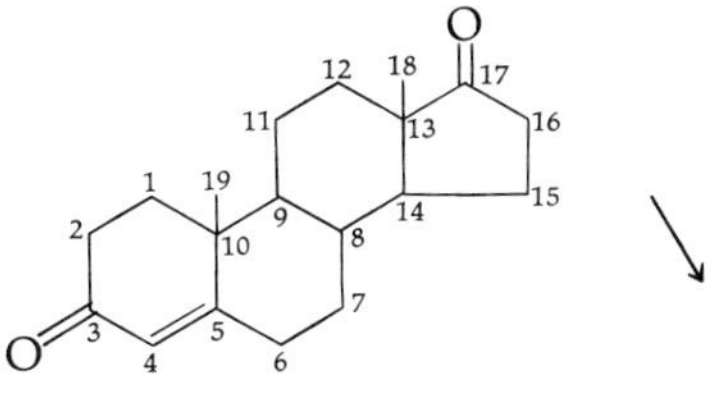

Androstenedione

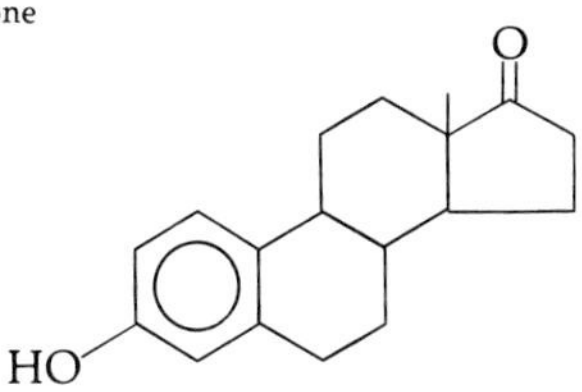

Oestrone

In an analogous way, oxidation of testosterone gives oestradiol.
[F:477]

527 Mixed function oxidases effect oxidations of two substrates with molecular oxygen. If RH is a suitable steroid substrate, the overall process could be represented as follows:

$$RH + O_2 + NADPH + H^+ \rightarrow$$
$$ROH + H_2O + NADP^+$$

[F:475]

## E Plasma lipoproteins

528 The 4 major groups are chylomicrons, very low density lipoproteins (VLDL), low density lipoproteins (LDL) and high density lipoproteins (HDL).
[I:217]

529 Albumin, containing adsorbed free fatty acid, is of high density and, if present, would sediment in the ultracentrifuge before lipoprotein fractions. The order of sedimentation of the lipoprotein fractions is (a) HDL, (b) LDL, (c) VLDL and (d) the chylomicron fraction.
[I:217]

530   Migrating more slowly than albumin, lipoprotein fractions will separate in order in $\alpha_1$ (or $\alpha$), $\alpha_2$ (or pre $\beta$) and $\beta$ bands, leaving chylomicrons at the origin. The relation of these with ultracentrifuge fractions is as follows: the $\alpha_1$ band is HDL, the $\alpha_2$ band is VLDL and the $\beta$ band is LDL.
[I:218]

531   All plasma lipoprotein fractions contain protein, phospholipid (PL), cholesterol (chol) and triglyceride (TG). Approximate particle sizes and compositions are tabulated.

| Fraction | Diameter | %Protein | %PL | %Chol | %TG |
|----------|----------|----------|-----|-------|-----|
| HDL | 10 nm | 50 | 25 | $1 + 16E^*$ | 8 |
| LDL | 20 nm | 25 | 21 | $8 + 38E^*$ | 8 |
| VLDL | 25–75 nm | 10 | 20 | $8 + 8E^*$ | 55 |
| Chylo-<br>  microns | 50–500 nm | 2 | 6 | $2 + 5E^*$ | 85 |

* E, cholesteryl esters.

[A:934]

532   Several apoprotein components of human plasma lipoproteins have been characterized (A-I, A-II, B, C-I, C-II, C-III, D and E). Apo-A-I and apo-A-II are the main protein components in HDL, apo-B is the main protein component in LDL and apo-B, apo-C and apo-E in VLDL. All the listed variants of apo-A, apo-B and apo-C are present in chylomicrons. The apo-C variants are important minor components of HDL. Specific functions have not been assigned to all the protein variants, but A-I and C-I are known activators of lecithin-cholesterol acyltransferase, whilst C-I and C-II are activators and C-III an inhibitor of (post-heparin) lipoprotein lipase.
[A:935]

533   Intestinal epithelial cells form chylomicrons in the presence of apo-B, and these are transported by the lymphatic system to enter the blood by means of the thoracic duct. Lipoprotein lipase can hydrolyse triglycerides in the chylomicrons and the fatty acid released is taken up by cells, including muscle and adipose tissue cells. The small remnant particles pass to the liver.
[A:941]

534   The major site of synthesis of lipid components is

the liver. Synthesized cholesterol and triglyceride combines with phospholipid, apo-B and other apoproteins to form VLDL. In peripheral tissues, triglycerides are hydrolysed by lipoprotein lipase and removed from the VLDL giving rise to intermediate density complexes IDL and then to LDL. This LDL is absorbed by cells, delivering cholesterol and triglyceride to the cells and partially suppressing the intracellular synthesis of these compounds.
[A:942]

535 HDL is formed in liver and in intestine. It has 3 main functions:
(a) it is the source of apo-C required to activate lipoprotein lipase and thus effect the hydrolysis of triglycerides in chylomicrons and in VLDL;
(b) it takes up cholesterol released by tissues and transports it to the liver;
(c) phospholipid in HDL, in the presence of LCAT enzyme, acts as the source of acyl groups for the formation of cholesteryl esters.
[A:948]

536 The following are rare inherited conditions:
(a) *lipoprotein lipase deficiency* leads to raised chylomicron levels, and is controlled by restricting dietary fat;
(b) *LCAT deficiency* leads to the accumulation of unesterified cholesterol in tissues and plasma;
(c) *HDL deficiency (Tangier disease)* gives rise to oily deposits of esterified cholesterol in tissues;
(d) *LDL deficiency* arises from lack of apo-B and, consequently, the apo-B-containing lipoproteins (chylomicrons, VLDL and LDL) are absent;
(e) *LDL receptor deficiency (or familial hypercholesterolaemia)* results in loss of control of (intracellular) cholesterol synthesis, with risk of cardiovascular disease.
[I:220]

537 Hyperlipoproteinaemia is likely in both conditions mentioned. Patients with high plasma lipid levels may develop isolated yellow fatty deposits (xanthomata) in their tissues. Very high plasma

VLDL and LDL levels are associated with atheromatosis.
[I:221]

538 Diabetes mellitus, hypothyroidism and the nephrotic syndrome are examples of conditions that can give rise to secondary hyperlipoproteinaemias, with raised plasma cholesterol and triglyceride levels.
[I:223]

# CONTROL OF METABOLISM

## A  Allosteric control

539 Control of the velocity of a key reaction in a pathway may be achieved by change in:
(a) the concentration of enzyme [by altering the rate of its synthesis (ribosomal or post-ribosomal), its catabolism, or its transport];
(b) the activity of the enzyme (by altering the concentration of activators or inhibitors at the site of the reaction);
(c) the concentration of substrate(s) and cofactor(s) in the neighbourhood of the active enzyme.
[B:291]

540 Allosteric control occurs when a molecule modifies the activity of a rate-limiting enzyme by attaching itself to the enzyme at a non-catalytic site. The effect is transferable from the allosteric site to the catalytic site because of the slight flexibility of the enzyme protein structure. The overall reaction rate of the pathway is altered and, concomitantly, the overall reaction rate of other connected pathways may also be altered.
[B:299]

541 Without detailed knowledge of the allosteric site, or of known allosteric effectors for the site, no specific prediction is possible. In very general terms, however, an allosteric effector may be a substrate or a product of a reaction in the pathway (including the allosterically modified reaction itself) or in a linked pathway. Cofactors, such as ATP, ADP, $NAD^+$ and NADH, exert allosteric effects in many pathways.
[B:296]

542 Multiple allosteric sites are known to exist in several types of enzyme molecule and these provide great scope for the delicate control of metabolic processes.
[G:221]

543 It is probable that all pathways are, directly or indirectly, under allosteric control. It is a major method of metabolic control.
[G:297]

544 A commonly quoted example of negative feedback allosteric influence is the inhibition of phosphofructokinase (and thus of glycolysis) by citrate.
[F:267]

545 Citrate may also be quoted as an example of an effector exerting a positive feedforward effect in that it can stimulate acetyl CoA carboxylase (and thus stimulate fatty acid synthesis).
[F:397]

546 Phosphofructokinase catalyses the reaction

Fructose 6-phosphate + ATP →

Fructose 1,6-diphosphate + ADP.

The substrate ATP *allosterically inhibits* the reaction and the product ADP *allosterically activates* the reaction if the concentration of the appropriate metabolite (ATP or ADP) is sufficiently elevated.
[A:191]

## B Regulation of substrate supply and compartmentation

547 The following sums of concentrations:
[ATP] + [ADP] + [AMP], [NAD$^+$] + [NADH], [NADP$^+$] + [NADPH], and [CoA] + [acyl CoA] tend to keep approximately constant values, at least over short periods of time. As these components are substrates (or products) for many pathways, and as their molarities are small, changes in concentration ratios, such as [NADH]/[NAD$^+$], can have striking effects on the velocities of many pathways.
[F:540]

548 Gluconeogenesis is occurring and we are dealing with the fasting state. In the fasting state, fatty acids from adipose tissue are undergoing β-oxidation in the liver and are converting available CoA into acetyl CoA. This severely reduces [CoA] and in consequence the rate of conversion of pyruvate to acetyl CoA is greatly reduced.
[F:545]

549 The ratio [lactate]/[pyruvate] would be an index of the cytosolic reducing state.
[D:298, 302]

550 Suitable small molecules can transfer reducing power across the mitochondrial membrane. Important examples are:
  (a) malate $^-$OOC.CHOH.CH$_2$.COO$^-$ transfers 2**H** into the cytosol,
  (b) glycerol 3–Ⓟ HOCH$_2$.**CHOH**.CH$_2$O-Ⓟ transfers 2**H** into mitochondria.
[D:249, 255]

551 Acetyl CoA cannot cross the mitochondrial membrane, but acetyl groups (**CH$_3$CO–**) can enter mitochondria as pyruvate **CH$_3$CO**.COO$^-$ and leave as citrate

HOOC.CH$_2$.C(COOH)OH.**CH$_2$.COO**$^-$.

[D:252]

## C Hormonal control

552 One type of hormone has molecules that become attached to specific cell membrane receptors and then modify the activity of adenyl cyclase present in the cell membrane. Peptide hormones act in this way.

The second type of hormone also has molecules that attach to specific cell receptors, but the molecules are then taken into the cells and transported by specific proteins to the nucleus where they influence transcription, usually by inducing the synthesis of a required enzyme. Steroid hormones act in this way.
[D:578]

553 The first type of hormone includes adrenaline, noradrenaline, thyroxine and the peptides insulin, glucogen, ACTH, vasopressin, parathyroid hormone, luteinizing hormone, thyroid-

stimulating hormone and melanocyte-stimulating hormone.
[D:578]

554 An adrenaline molecule becomes attached at a specific receptor site on a cell surface and stimulates adenyl cyclase to convert some ATP into cAMP. Each molecule of cAMP in the cell can allosterically activate a cAMP-dependent kinase by combining with, and effecting the removal of, (inhibiting) regulatory subunits in the inactive enzyme. The catalytic effect is then amplified, because the active cAMP-dependent kinase catalyses the phosphorylation of inactive phosphorylase $b$ kinase. The active phosphorylase $b$ kinase produced in its turn catalyses the phosphorylation of inactive phosphorylase $b$ to give the active enzyme phosphorylase $a$. The latter catalyses the phosphorolysis of unbranched chains in glycogen. The amplification system is called an enzyme cascade.
[C:171]

555 Both adrenaline and ACTH can attach to receptors on adipocytes and insulin can block these receptors. Bound adrenaline and ACTH then stimulate membrane-bound adenyl cyclase to convert ATP into cAMP. The cAMP allosterically activates a cAMP-dependent kinase which in turn phosphorylates, and thus activates, inactive triglyceride lipase. The active lipase catalyses the hydrolysis of triglycerides.
[C:225]

556 In muscle, for example, a specific phosphodiesterase enzyme is known that catalyses the rapid removal of cAMP by the reaction:

$$cAMP + H_2O \rightarrow AMP$$

Phosphatases are also present and these may be able to hydrolyse phosphorylated forms of enzymes.
[F:842]

557 Stimulation of a pathway $A \xrightarrow{\text{route 1}} Z$ would have little value if the effective reverse pathway $Z \xrightarrow{\text{route 2}} A$ were to undergo comparable acceleration. In the breakdown of the straight chains in

glycogen, a phosphorylated enzyme (phosphorylase *a*) is the active catalyst, whilst, in the synthesis of the unbranched chains, a non-phosphorylated synthetase is the active catalyst and its phosphorylated form is inactive. Thus, the hormonal mechanism that switches on breakdown, switches off synthesis and *vice versa*.
[F:372]

558 Many processes are known to be stimulated by cAMP. These include glycogen breakdown, gluconeogenesis, lipolysis, ketogenesis, insulin release from the pancreas, contraction of cardiac muscle, amylase release from the parotid and HCl release from the gastric mucosa.
[F:843]

559 The hormones of the adenohypophysis have the following main functions:
(a) somatotropin (SH, growth hormone) stimulates protein synthesis,
(b) adrenocorticotropin (ACTH) stimulates corticosteroid synthesis and release from the adrenal cortex,
(c) thyrotropin (TSH) stimulates the thyroid to produce and release $T_3$, tri-iodothyronine and $T_4$, thyroxine (tetraiodothyronine),
(d) melanotropin ($\beta$-MSH) stimulates melanocytes,
(e) prolactin (PRL) stimulates the mammary gland,
(f) follicle-stimulating hormone (FSH) stimulates both maturation of the follicle and spermatogenesis,
(g) luteinizing hormone (LH) stimulates ovulation, formation of the corpus luteum and progesterone secretion, and also stimulates testosterone production.
[D:586]

560 Hypothalamic regulatory hormones so far characterized are all small peptides, the very few amino acid residues sometimes possessing minor chemical modifications (such as cyclization of an N-terminal glutamic acid or the replacement of a terminal –COOH by a –$CONH_2$ group).
[D:585]

561 Release of the hormone somatotropin (SH) is stimulated by the somatotropin-releasing hor-

mone (SRH) and inhibited by the hormone somatostatin (SRIH). [Somatostatin contains 14 amino acid residues (a comparatively large number for a hypothalamic regulatory hormone).]
[D:585]

562 Corticotropin-releasing hormone (CRH) from the hypothalamus stimulates the secretion of ACTH from the adenohypophysis and the ACTH, in turn, stimulates the synthesis of glucocorticoids from cholesterol in the adrenal cortex. High cortisol production inhibits the release of both CRH and ACTH, and high ACTH levels also inhibit CRH release. Low plasma cortisol levels stimulate CRH production. A specific $\alpha$-globulin, transcortin, transports glucocorticoids in plasma.
[D:599]

# MINERAL METABOLISM

## A  Sodium, potassium and water

563 The major cation and anion of extracellular fluid are $Na^+$ and $Cl^-$ respectively, whilst the major cations of intracellular fluid are first $K^+$, with the highest concentration, followed by $Mg^{2+}$. (In this area, milliequivalents per litre is often a more appropriate method of expressing an ion concentration than mmol per litre.)
[G:1014]

564 The term 'sodium pump' refers to the mechanism that actively transports $Na^+$ ions from the interior to the exterior of a cell with the concomitant transfer of $K^+$ ions from the exterior to the interior fluid. Two $K^+$ ions are transferred for each 3 $Na^+$ ions. The process is energy-requiring and utilizes an $Mg^{2+}$, $Na^+$, $K^+$ dependent ATPase. This enzyme is inhibited by the digitalis derivative ouabain.
[F:862]

565 Excess of $Na^+$ is excreted largely by the kidneys although some is lost in sweat. Faecal losses are small except with diarrhoea. Severe diarrhoea, arising from cholera or other infections, can result in considerable loss of water and of $Na^+$ and $Cl^-$ ions.
[I:31]

566 Virtually all the $K^+$ and approximately 80% of the $Na^+$ and of the $Cl^-$ are reabsorbed from the glomerular filtrate in the proximal tubule. In the loop of Henle, an active countercurrent process occurs that, in effect, pumps $Na^+$ and $Cl^-$ from the ascending back to the descending limb by a process of reabsorption from the ascending and secretion into the descending limb. In the distal tubule, under the action of aldosterone, $Na^+$ is reabsorbed from, whilst $K^+$ and $H^+$ are secreted into, the tubular lumen. Reabsorption of water then occurs, mainly from the collecting duct, as the latter passes through the hyperosmolal region of the medulla containing the lower part of the loop of Henle. This water reabsorption requires the presence of vasopressin (antidiuretic hormone ADH). Any reabsorption of $Cl^-$ is passive.
[I:3, 6, 10]

567 Aldosterone stimulates the distal tubule's sodium pump so that $Na^+$ ions are reabsorbed and $K^+$ and/or $H^+$ ions are excreted. Increase in $Na^+$ reabsorption will be offset usually by increased retention of water with restoration of circulating volume. Reduction in extracellular fluid volume acts at the juxtaglomerulus to stimulate the renin–angiotensin–aldosterone hormone system and aldosterone secretion is enhanced.
[A:1195]

568 Aldosterone influences $Na^+/K^+$ transport and possibly $Na^+/H^+$ transport across all cell membranes. In relation to excretion, it has an $Na^+$ ion retaining effect on the sweat glands and on the gastrointestinal tract as well as on the kidneys.
[C:491]

569 The action of vasopressin (ADH) is to increase the permeability to water of the cells lining the renal collecting ducts. Passive reabsorption of water occurs into the hyperosmolal region (renal medulla). This controlled water reabsorption leads to the excretion of an appropriately concentrated urine.
[A:1189]

570 The $K^+$ ion is predominantly an intracellular ion. It forms a complex with pyruvate kinase, essential for catalytic activity. Thus it is necessary for the operation of the glycolytic pathway. The $K^+$ ion may be required by some non-glycolytic en-

zymes. Maintenance of the correct (low) extra-cellular $K^+$ concentration is important for normal cardiac action and neuromuscular activity.
[A:1200]

571  Hyperkalaemia (raised plasma $[K^+]$) can arise from reduced adrenal activity, kidney disease, cell breakdown in wasting diseases, and in diabetic acidosis. It may be countered if cellular uptake of $K^+$ can be raised by stimulating glycolysis with insulin or, where an acidosis exists, by taking steps to correct the plasma pH.
[A:1200]

572  Hypokalaemia may result from:
(a) $K^+$ loss from the gut (vomiting, diarrhoea, intestinal fistula leakage);
(b) $K^+$ loss in urine
(i) from distal tubules by increased $Na^+/K^+$ exchange (e.g. because of high aldosterone level) or reduced $Na^+/H^+$ exchange,
(ii) from proximal tubules, e.g. in the Fanconi syndrome or in renal tubular failure;
(c) reduced $K^+$ intake in chronic starvation;
(d) increased glycolysis due to insulin therapy with a large carbohydrate intake.
In pyloric stenosis, with alkalosis, $K^+$ ions are lost from the plasma into cells, into intestinal fluid, and into urine.
[A:1201]

# B  Calcium, magnesium and phosphorus

573  An average normal 70 kg man contains about 1200 g of calcium, 25 g of magnesium and 700 g of phosphorus.
[B:797]

574  Calcium, present mainly as a type of phosphate, is the main metal component of the mineral phase of hard tissues, e.g. bone and teeth. Calcium ions also exert essential functions in blood clotting, muscular contraction, the transmission of nerve impulses, neuromuscular irritability and in the function and maintenance of cell membranes. Calcium ions influence some intracellular processes through their effect on the conformation of the protein calmodulin.
[D:22, 593]

575 Magnesium is an essential cofactor for many enzymes, particularly for reactions involving ATP (where an ATP–Mg complex is the substrate). Other functions of magnesium in man appear to relate to the activities of calcium ions. For example:

(a) over 50% of the body's magnesium is present in bone and $Mg^{2+}$ ions move into and out of bone along with $Ca^{2+}$ ions.

(b) $Mg^{2+}$ ions, like $Ca^{2+}$ ions influence neuromuscular irritability, one consequence being that reduced plasma $Mg^{2+}$ concentration, as well as reduced plasma $Ca^{2+}$ concentration, can lead to tetany.

Intracellular magnesium concentrations are always much higher than those in extracellular fluid in contrast to the distribution of calcium. This suggests that the magnesium/calcium relationship might be analogous in some respects to the potassium/sodium relationship.
[D:22]

576 About 80% of the phosphorus in the body is present, in hydroxyapatite, in bone and teeth. The remainder is present in $H_2PO_4^-$ and $HPO_4^{2-}$ ions, in nucleic acids and in the phosphate esters, acid anhydrides and amides that participate in a wide variety of metabolic pathways.
[C:556]

577 Calcium in plasma is present in

(a) free $Ca^{2+}$ ions;

(b) protein-bound calcium ($Ca^{2+}$ bound mainly to albumin);

(c) calcium ions complexed to small anions such as lactate, citrate and phosphate.

Total plasma $Ca^{2+}$ levels are normally within a narrow range of $2.5 \pm 0.25$ mmol/l ($10 \pm 1$ g/dl). The proportion present as free $Ca^{2+}$ ions depends on blood pH and other factors, but is normally about 45–50%. Slightly less is present in the protein-bound form, with less than 10% complexed to small anions. The only physiologically active form is free ionic calcium $Ca^{2+}$. Release of $Ca^{2+}$ from the protein-bound form takes place very slowly should the $Ca^{2+}$ concentration fall.
[I:232]

578 Increased neuromuscular activity occurs in hypocalcaemia and this can give rise to a state of

muscle spasms (tetany). The abnormal neuromuscular effects include increased cardiac contractibility and retarded relaxation. Death may follow from cardiac or respiratory failure.

Chronic mild hypocalcaemia can lead to the formation of cataracts and also to psychiatric symptoms.
[I:237]

579　High plasma $Ca^{2+}$ ion concentration depresses neuromuscular excitability. One consequence can be constipation and abdominal pain. Severe hypercalcaemia may produce cardiac arrest.

Prolonged mild hypercalcaemia carries the risk of kidney damage due to calcification of renal tubules or the formation of stones.
[I:236]

580　Plasma $Ca^{2+}$ concentration is controlled, normally within a narrow band, by the action of the two polypeptides, parathyroid hormone (parathormone, PTH), secreted by the parathyroid glands, and calcitonin, formed in the thyroid. Vitamin D acts to increase intestinal absorption of calcium and influences the actions of the polypeptide hormones.
[A:1000, 1007, 989]

581　The term 'vitamin D' is used to cover a group of closely related substances that effectively exert an antirachitic action. The substances include ergocalciferol (vitamin $D_2$), cholecalciferol (vitamin $D_3$), their respective 25-hydroxy derivatives and 1,25-dihydroxycholecalciferol.
[A:989]

582　Vitamin D can be obtained by intestinal absorption of fat-soluble dietary vitamin $D_3$ or $D_2$ and it can also be obtained by the epidermal formation of vitamin $D_3$ (cholecalciferol) from its isomer 7-dehydrocholesterol. The latter compound is a precursor in the synthesis of cholesterol. The energy of ultraviolet radiation in sunlight acts on 7-dehydrocholesterol, transferring an H atom from carbon atom 19 to carbon atom 9. This breaks the bond between carbon atoms 9 and 10 and thus opens ring B. Although cholecalciferol does not contain the steroid ring system, the numbering of the carbon atoms is unaltered.
[C:116]

583 A microsomal mixed function oxidase in liver cells converts cholecalciferol into its 25-hydroxy derivative which passes to the kidney. A mitochondrial mixed function oxidase in kidney cells effects further hydroxylation to give the active metabolite 1,25-dihydroxycholecalciferol (1,25-DHCC).
[C:117]

584 Low serum phosphate concentration stimulates the 1-hydroxylation reaction directly and low serum calcium concentration stimulates the parathyroid glands to secrete PTH, which then stimulates the 1-hydroxylation reaction. In normal and hypercalcaemic conditions, hydroxylation can occur at carbon atom 24.
[C:118]

585 1,25-DHCC enters the intestinal wall cells and unmasks a specific gene which transcribes a specific mRNA. This mRNA codes for a calcium-binding protein that effects the transport of $Ca^{2+}$ ions across the intestinal wall.
[C:119]

586 PTH molecules (from different species) all consist of a single polypeptide chain of 84 amino acid residues, but with varying amino acid compositions. The biological activity resides in the sequence of 34 amino acid residues from the N-terminal end.

Calcitonin (from several species) always consists of a single chain of 32 amino acid residues, but with variation in the amino acid composition for each species. All forms contain a disulphide bridge between two cysteine residues, separated from each other by five amino acid residues.
[G:1211, 1216]

587 PTH is secreted by the parathyroid glands at a rate that increases as the plasma $Ca^{2+}$ concentration drops (and *vice versa*). It acts to raise the plasma $[Ca^{2+}]$ and then maintain it at the normal level. It acts by:
(a) increasing the osteolytic activity of existing osteoclasts and then increasing the rate of conversion of undifferentiated cells and of osteocytes into osteoclasts with simultaneous delay in the conversion of osteoclasts into osteoblasts (bone-forming cells), all these

processes increasing the resorption of bone;

(b) increasing the renal tubular reabsorption of $Ca^{2+}$ (the reabsorption of phosphate is diminished);

(c) stimulating the synthesis of 1,25-DHCC (in the kidney), which then acts to increase $Ca^{2+}$ absorption from the gastrointestinal tract and to synergistically augment the stimulating effect of PTH on osteoclastic activity.

[A:996, 1000]

588 Calcitonin, formed in the parafollicular cells of the thyroid, acts to reduce high plasma $Ca^{2+}$ concentration and to maintain the normal plasma level. Its important activity is to inhibit bone resorption by controlling the activity and number of osteoclasts. There is some evidence that this inhibiting effect is accelerated by the presence of 1,25-DHCC. The hypocalcaemic hypophosphataemic effects of calcitonin are most apparent in children and under circumstances where rapid remodelling of bone is taking place.

It has also been reported that calcitonin increases renal excretion of $Ca^{2+}$ and phosphate and inhibits renal synthesis of 1,25-DHCC.

[A:1007]

589 It would be expected that at least some of the activities of PTH and calcitonin would operate through the formation of cAMP in the target cells.

[G:1214, 1218]

590 Lack of vitamin D in rickets severely restricts intestinal $Ca^{2+}$ absorption and, as a consequence, plasma $Ca^{2+}$ concentration falls. The poor rate of production of 1,25-DHCC is partly offset by enhanced secretion of PTH. The high plasma PTH level produces a rate of bone resorption that retards bone development.

[A:1015]

591 A low plasma phosphate concentration stimulates kidney cells to synthesize 1,25-DHCC and this compound, by augmenting bone resorption and intestinal absorption, can effect the increase of both plasma $Ca^{2+}$ and plasma phosphate concentrations. Thus the hypophosphataemia is corrected whilst, because the retaining influence of

PTH will be absent, the excess of $Ca^{2+}$ ions will be excreted in the urine.
[A:997]

## C  Iron and other transition metal ions

592  An adult male contains about 4 g of iron. About 70% is present in haemoglobin in erythrocytes, about 20% is stored in the reticuloendothelial system as ferritin (water-soluble) and as haemosiderin (granules in liver cells) and about 10% is present as myoglobin, cytochromes and iron-containing enzymes. About 0.1% is bound to protein [transferrin] circulating in the plasma.
[B:801]

593  An adult male, who is neither a blood donor nor has lost blood through haemorrhage, is almost a closed system for iron as virtually no iron is excreted in urine, bile or sweat. The only loss would arise from desquamated cells and this loss (about 1 mg per day) occurs mainly into the intestinal tract. In normal women, of child-bearing age, the average daily loss will be about 2 mg in total. The body's iron content is controlled at the absorption stage. Iron can only cross cell membranes in the Fe(II) state, but in storage (as ferritin and haemosiderin) and in transport (in association with transferrin) it is present in the Fe(III) state. Only a small proportion of dietary iron is absorbed, the amount being dependent on requirement and also on the form in which the iron is ingested.
[A:682, 679]

594  Three important ways in which iron-deficiency anaemia may arise are:
(a) loss of iron in women due to menstruation, pregnancy and lactation;
(b) blood loss in the gastrointestinal tract due to ulceration or malignancy;
(c) deficient iron absorption, possibly because of a diet rich in phosphates or other iron-binding components or containing oxidants.
[A:688]

595  Iron overload may arise from multiple blood transfusions (without concomitant loss of blood) or by excessive intake and absorption of dietary iron (e.g. consumption of alcoholic drinks with

high iron content). Such unusual circumstances need to last several years to produce a clinically important iron overload. Two recognized syndromes of iron overload are:

(a) ideopathic haemochromatosis, probably caused by an inherited defect that increases intestinal absorption of iron from a normal diet,

(b) Bantu siderosis, probably caused by the consumption, over many years, of native beer brewed in iron vessels.

[I:389]

596 Copper is a trace element in man, an adult containing about 0.1–0.15 g. It is a component of several enzymes, including cytochrome oxidase, monoamine oxidase, ascorbic acid oxidase, lysyl oxidase, aminolaevulinate dehydrase and tyrosinase. It is concerned with a variety of functions including mitochondrial energy production, the cross-linking of collagen, the stimulation of erythropoiesis and melanin formation.

[A:1019]

597 Wilson's disease, or hepatolenticular degeneration, is a rare inherited disturbance of copper metabolism, characterized by abnormally large amounts of copper complexes in liver and brain and also in kidney and eye. Biliary excretion of $Cu^{2+}$ is decreased, but renal excretion is enhanced. In serum, copper is normally complexed to a protein as ceruloplasmin and urinary excretion is small. Ceruloplasmin is deficient in Wilson's disease so that urinary excretion of copper increases.

Tissue damage is caused by the abnormal deposits of copper compounds.

[A:1021]

598 Zinc ions form an essential trace component of the diet as they are required for the activity of enzymes concerned in the metabolism of proteins, carbohydrates, lipids and nucleic acids. Examples of zinc-containing enzymes are carbonic anhydrase, carboxypeptidase, alcohol and lactate dehydrogenase, alkaline phosphatase and DNA and RNA polymerases.

[A:1022]

599 Cobalt is bonded into a porphyrin-like structure

as part of cobamide coenzyme derived from vitamin $B_{12}$. The coenzyme is concerned in the conversion of ribonucleotides into deoxyribonucleotides for DNA synthesis and in the transfer of one-carbon units. Vitamin $B_{12}$ acts synergistically with folic acid in haemopoiesis. Its deficiency causes pernicious anaemia.
[A:1064]

## D  Anions (excluding phosphate)

600  Soluble fluoride is rapidly absorbed in the gut and readily excreted in urine. Some fluoride enters teeth and bone, producing fluoroapatite, which is a more resistant structure than the hydroxyapatite of teeth and bone. This reduces the erosion of teeth. The addition of fluoride ions to public drinking water supplies is carried out in many localities where the concentration of natural fluoride is below 1 part per million of water. Fluoride ions in much higher concentration inhibit several metabolic enzymes and have toxic effects.
[H:240]

601  The chloride shift (into erythrocytes) forms such an example. Metabolically produced carbon dioxide enters the erythrocyte and is rapidly converted into carbonic acid by carbonic anhydrase. Ionization of the carbonic acid is enhanced by the acceptance of hydrogen ions by $HbO_2$, leaving bicarbonate ions that diffuse out of the cell. The electrical charge displacement that the migration of bicarbonate would produce is offset by the diffusion of chloride ions into the erythrocytes.
[C:522]

602  In the parietal cells of the gastric mucosa, carbonic anhydrase produces carbonic acid from carbon dioxide and water. The carbonic acid ionizes and the hydrogen ions diffuse out of these cells. Chloride ions in the tissue fluid are transported into the stomach along with the hydrogen ions. The process of secretion of hydrochloric acid into the stomach occurs under the influence of the peptide hormone gastrin.
[C:527]

603  The rich sources of iodide are sea foods and sodium chloride containing iodide (artificially

added). The iodide ions are readily absorbed intestinally and are actively taken up by the thyroid to produce thyroglobulin. It is the break-down of thyroglobulin that releases thyroid hormones.
[B:711]

604 The thyroid hormones accelerate many metabolic processes. They are required at an appropriate level for the maintenance of normal heart rate and cardiac output and for normal growth, mental development and sexual development. In hyperthyroid states, metabolic processes are speeded up, whilst in hypothyroid states they are slowed down.
[C:468]

605 Sulphate is converted into 'active sulphate' by:
(a) reaction with ATP in the presence of a sulphurylase enzyme and $Mg^{2+}$ to give AMP sulphate, which contains the group $-O-PO_2^--O-SO_3^-$;
(b) further phosphorylation of AMP sulphate in the 3'-position of the ribose ring (by means of ATP in the presence of $Mg^{2+}$ and a kinase). The resulting reactive metabolite 3'-phospho - adenosine - 5' - phosphosulphate (PAPS) can readily transfer sulphate to other suitable metabolites in the presence of the appropriate transferase.
[A:624]

# LIVER AND BILE

## A  Functions of the liver and some inborn errors

606 All nutrients from the gut, except lipids, travel through the portal vein to the liver before entering the systemic circulation. The liver is extremely active in all main areas of general metabolism. Major pathways that operate include:
(a) glycogen synthesis and breakdown,
(b) gluconeogenesis and glycolysis,
(c) fatty acid synthesis and oxidation,
(d) triglyceride synthesis and breakdown,
(e) phospholipid synthesis and breakdown,
(f) ketone body formation,
(g) cholesterol synthesis,
(h) formation of 25-hydroxycholecalciferol,

(i) bile acid synthesis,
(j) lipoprotein synthesis (and probable breakdown of HDL),
(k) amino acid synthesis and catabolism,
(l) urea formation.
[I:285]

607 The liver possesses detoxifying, excretory, storage and reticuloendothelial functions. Thus, steroids and drugs are modified and usually made more suitable for excretion (either in bile or urine), whilst iron, glycogen and some vitamins are stored.
[A:1275]

608 The 2 inborn errors presented in this way are likely to be galactosaemia and glycogen storage disease type I (often called von Gierke's disease). Galactosaemia, however, can be detected by screening tests when the infant is born and is even detectable before birth.
[I:202]

609 Galactokinase catalyses the ATP-mediated conversion of D-galactose into D-galactose 1-phosphate. The latter compound reacts with UDP-glucose by a reversible reaction to give glucose 1-phosphate and UDP-galactose. This reaction is catalysed by a transferase, hexose 1-phosphate uridyl transferase, which is almost absent from the liver and erythrocytes of galactosaemic infants. Finally, an epimerase can catalyse (reversibly) the conversion of UDP-galactose into UDP-glucose. The condition is rare and is inherited as an autosomal recessive trait arising from a structural gene mutation.
[A:311]

610 There is variation in the clinical severity of the condition, but patients with galactosaemia usually lose weight, experience vomiting and diarrhoea and suffer liver damage. They also show evidence of mental retardation, renal tubular damage and they tend to develop cataracts. There is galactosuria and a secondary aminoaciduria. Clinical effects are caused by the buildup of galactose 1-phosphate and metabolites of this compound, including galactitol. The condition is treated by the elimination of lactose and

galactose from the diet as soon as possible. The infant is still capable of synthesizing galacto-lipids etc. because of the reversible character of the epimerization:

UDP-galactose $\rightleftharpoons$ UDP-glucose

[A:312]

611 The defective enzyme in von Gierke's disease is glucose 6-phosphatase. Lack of this enzyme results in fasting hypoglycaemia because the formation of glucose from both glycogenolysis and gluconeogenesis is blocked. Indirect consequences are the development of lactic acidosis from the raised pyruvic acid levels, and the development of ketosis and hyperlipoproteinaemia as a consequence of low glucose levels.
[A:285, 284]

612 In von Gierke's disease the child exhibits loss of weight and has vomiting episodes. The enlarged liver contains cells loaded with glycogen. Fasting hypoglycaemia, lactic acidosis, ketosis, and hyperlipoproteinaemia occur and competition with lactate in renal excretion can produce a uricaemia. Treatment must be based on the maintenance of normal plasma glucose concentrations by frequent suitable feeding.
[A:286]

613 In addition to von Gierke's disease (type I), there are several other rare inherited diseases that are characterized by the deposition in the liver or other tissues of abnormally large glycogen stores. The glycogen in most of these diseases is normal in structure, but 2 diseases have glycogens of an abnormal molecular structure. The diseases arise from defects in enzymes concerned with glycogen metabolism.
[A:284, 286]

## B  Bile pigment metabolism

614 When erythrocytes are broken down, the haemoglobin divides into globin, which enters the protein pool, and haem. The haem is then split by a microsomal enzyme system, releasing the iron, oxidizing one of the bridging CH groups to carbon monoxide, and forming the

blue-green pigment biliverdin. In man, biliverdin is immediately reduced by NADPH, in the presence of a reductase, to give the related, orange-yellow, insoluble bilirubin of structure

where $M$ = methyl, $V$ = vinyl and $P$ = propionic acid (2-carboxy-ethyl-).
[C:317]

615    The newly formed bilirubin binds to serum albumin and is carried to liver cells. The complex dissociates and the bilirubin enters the liver cell, being accepted by 2 carrier proteins Y and Z. The affinity of bilirubin for Y is greater than that for Z, but the binding capacity of Z is greater than that of Y. Ligandin (or Y protein) is slowly formed in the newborn and the full complement of Y is not formed until an infant is several weeks old. Intracellular bilirubin is conveyed to the smooth endoplasmic reticulum, where it undergoes reaction with 2 molecules of UDP-glucuronic acid at the 2 propionic acid groups. The bilirubin diglucuronide formed is called conjugated bilirubin. The transferase enzyme is present in barely sufficient amounts at birth and requires a few weeks to develop to adult levels.
[A:785]

616    The highly water-soluble bilirubin diglucuronide is secreted into the bile canaliculi and enters the bile for excretion. The conjugated bilirubin is not reabsorbed intestinally, but bacteria in the colon effect its stepwise reduction (with some deconjugation) to form a set of colourless compounds generally known as urobilinogen. Most of these materials are oxidized to give orange-yellow products, called collectively faecal urobilin (a main component is called stercobilin), and then eliminated in faeces. A small proportion of urobilinogen is reabsorbed and this is taken in the portal circulation to the liver, where most is re-excreted in the bile. About 4 mg per day is normally excreted in urine.
[A:790]

617 Bilirubin diglucuronide (conjugated bilirubin) in solution (e.g., in serum) will react with a solution of diazotized sulphanilic acid (van den Bergh's reagent) to yield an azobilirubin. The intensity of colour produced by this red dye gives a measure of the bilirubin concentration. This is called a direct van den Bergh reaction. Bilirubin itself, and in association with albumin, will only give the red colour if either ethanol or methanol is present in adequate concentration. This latter is called an indirect van den Bergh reaction. These reactions are of value in detecting and determining bilirubin in serum. A positive indirect reaction and a negative direct reaction indicates a prehepatic (e.g., a haemolytic) jaundice.
[A:794]

618 Neither unconjugated nor conjugated bilirubin is present in normal urine. Unconjugated bilirubin, being associated with albumin, cannot pass the glomerular filter, whilst conjugated bilirubin is secreted into the bile and is not reabsorbed. Urobilinogen is present in traces in fresh normal urine and is often detectable by simple tests, particularly if the urine happens to be concentrated. The urobilinogen is converted into coloured urobilin (virtually absent from fresh urine) if the urine is allowed to stand.
[I:287]

## C Jaundice

619 Jaundice is a physical condition in which an abnormally high plasma concentration of bilirubin (unconjugated and/or conjugated) produces a yellow staining of surface tissues (including the sclerae of the eyes) that is recognizable by the clinician. Normal plasma contains unconjugated bilirubin, the concentration not exceeding 15 $\mu$mol/l. About 3 times the normal concentration will produce recognizable tissue staining and 5 times gives rise to obvious jaundice.
[A:797]

620 Prehepatic jaundice exists when the plasma concentration of unconjugated bilirubin is greatly increased. Posthepatic jaundice is caused by biliary obstruction, the resulting bilirubinaemia being caused mainly by conjugated bilirubin entering the plasma. Hepatic jaundice is caused by

the necrosis of liver cells. Both the conjugation process and transport of conjugated bilirubin into the bile is inhibited, so that the bilirubinaemia is due to both unconjugated and conjugated bilirubin. The jaundice of the Dubin–Johnson syndrome does not lend itself to ready classification. [G:990]

621 The liver normally conjugates about 0.3 g of bilirubin daily, but it is capable of dealing with much larger quantities. However, if the concentration of circulating bilirubin–albumin complex is sufficiently high, the normal liver will be unable to cope with such high loads and jaundice will result. A liver that is defective in processing bilirubin could be overwhelmed by even the normal plasma level of unconjugated bilirubin.

A high plasma concentration of unconjugated bilirubin can arise from increased bilirubin production, e.g. from excessive haemolysis (including the breakdown of immature erythrocytes) or from the absorption of a large haematoma. However, the unconjugated hyperbilirubinaemia might be the result of:
(a) an abnormally low rate of uptake of bilirubin by liver cells or of its transport within the cells, as exemplified by the congenital Gilbert's disease;
(b) disturbances of conjugation within hepatocytes, as exemplified by an immature transferase enzyme in neonates, congenital transferase deficiency (Crigler–Najjar syndrome), and transferase inhibition by drugs.
[A:798]

622 (a) Posthepatic jaundice can arise from biliary obstruction caused by gall stones, carcinoma of the head of the pancreas, constriction of the duct by external pressure (e.g., by a tumour) and fibrosis of the duct.
(b) Hepatic jaundice can arise from cell damage caused by hepatitis, drugs, and toxins and from intrahepatic cholestasis resulting from chronic hepatitis, drugs, autoimmune biliary cirrhosis and (sometimes) cirrhosis.
All these conditions produce a rise in the concentration of conjugated bilirubin in the liver and this soluble pigment enters the plasma. In addition, liver cell damage will directly reduce the ability of the liver to process unconjugated biliru-

bin so that plasma levels of both conjugated and unconjugated bilirubin will rise when liver damage is pronounced. Finally, it is possible that a high concentration of conjugated bilirubin in the liver cells might itself inhibit the conjugating process and so some increase in unconjugated bilirubin might accompany the conjugated bilirubinaemia for this reason.
[A:799]

623 If biliary obstruction is virtually complete then almost no conjugated bilirubin is secreted in the bile and, as a result, hardly any urobilinogen forms in the gut. Consequently:
(a) conjugated bilirubin enters the plasma and is excreted in the urine;
(b) urobilinogen is not excreted in the urine.
[I:294]

624 Jaundice is very common in the neonate for the following reasons:
(a) UDP-glucuronyl transferase is not synthesized in sufficient amounts, so that conjugating power is low;
(b) synthesis of Y protein may be deficient, thus limiting the capacity of cells to transfer unconjugated bilirubin to the microsomal sites of conjugation;
(c) haematomas produced by injuries at birth may increase the load of unconjugated bilirubin.
More severe jaundice will be caused by disease such as haemolytic disease of the newborn, neonatal hepatitis and abnormalities of the biliary tract.
[A:800]

625 Many drugs and toxins have damaging effects on the liver and jaundice may result, but drug-induced jaundice may develop without liver damage. Any such effects of a drug become of particular consequence if the patient has special sensitivity or if the drug is prescribed over long periods of time.
    The newborn, with immature bilirubin processing, are at risk because some drugs can produce severe jaundice that can lead to kernicterus. Novobiocin, which inhibits the glucuronyl transferase system, salicylates and sulphonamides,

which displace bilirubin from albumin, and drugs producing haemolysis must be avoided.

Several toxic materials, e.g. carbon tetrachloride, and some drugs, particularly in overdose, produce liver necrosis. Other drugs (iproniazid, isoniazid) can produce an effect that resembles viral hepatitis, whilst phenothiazines, such as chlorpromazine, and $17\alpha$-alkyl steroid derivatives, present in oral contraceptives, can produce a posthepatic jaundice.
[I:297]

## D  Biochemical tests and liver disease

626   The BSP test is a very sensitive indicator of any diminution in the liver's excretory function. A solution of the dye BSP is administered intravenously and the proportion remaining in the circulation after 45 minutes is determined. As normal liver cells rapidly take up the dye, conjugate it and excrete it into the bile, less than 5% should remain in the circulation after 45 minutes. The test is applied if other tests show no dysfunction. It is superfluous if the patient is jaundiced.
[A:1277]

627   Some plasma proteins, synthesized in the liver, can act as an index of liver synthesizing function. Albumin is rapidly turned over and reduced synthesis is soon reflected in low serum values. Other proteins synthesized in the liver include prothrombin and other blood clotting factors. The activities of some of these factors is measured by the one-stage prothrombin time. Vitamin K is required for the synthesis of these factors. Consequently, low measured activities may be due to liver cell damage, reducing synthesis even in the presence of adequate vitamin K, or to cholestasis, reducing the flow of bile and restricting the intestinal absorption of the fat-soluble vitamin K. Shortening of the prothrombic time after injection of vitamin K indicates that cholestasis is present, whilst no change in prothrombic time indicates liver cell damage.
[I:292]

628   Alanine transaminase (ALT), aspartate transaminase (AST) and isocitrate dehydrogenase are enzymes with serum activities that tend to increase with liver cell damage. Isocitrate DH is

more specific for liver damage than ALT or AST, but the latter enzyme activities are more commonly measured.
[I:291]

629 Alkaline phosphatase (ALKP), γ-glutamyl-transferase, 5'-nucleotidase and leucine amino-peptidase have activities in serum that increase when cholestasis is present. ALKP, which is present in the cells lining the sinusoids and the bile canaliculi, is the serum activity commonly measured. ALKP in serum is also derived from bone, kidney and other tissues.
[I:291]

630 No information concerning liver disease is available from β-globulin fractions, but γ-globulin levels increase in cirrhosis and in chronic liver disease. Determinations of particular immuno-globulins can sometimes be of value (IgM is raised in autoimmune biliary cirrhosis).
[I:295, 293]

631 When jaundice is observed, the urine should be examined for bilirubin (conjugated) and uro-bilinogen (or urobilin). Plasma (or serum) bilirubin levels should be measured and suitable serum enzyme assays (e.g., ALT, AST and ALKP) should be made. The information obtained will greatly aid a rapid diagnosis in many patients.
[A:804, 806]

632 If acute or chronic hepatitis is expected the B antigen (HBsAg) should be measured.
α-Fetoprotein is a protein present in fetal serum, but it becomes detectable in adults with primary hepatocellular carcinoma.
The detection of circulatory antibodies, such as mitochondrial antibody (M), can be of value. Thus M antibody is present in most cases of autoimmune biliary cirrhosis, but is usually absent in cases of extrahepatic biliary obstruction.
[I:293]

# E  Bile

633 The liver normally produces about 1.5 litres of bile daily containing bile salts, cholesterol, phospholipids, bicarbonate, chloride, sodium and

other ions, conjugated bilirubin and some conjugated or unconjugated waste or toxic materials.
[I:299]

634 The gall bladder both stores and concentrates bile. Active reabsorption of sodium, chloride and bicarbonate ions occurs together with absorption of about 90% of the water content.
[I:299]

635 Bile has the following physiological properties:
(a) it is basic and acts to neutralize acid from the stomach;
(b) it has a powerful emulsifying action, vital for the intestinal absorption of fats, fatty acids and fat-soluble vitamins;
(c) cholesterol is incorporated into a lecithin-bile salt micelle that is transported into the intestine;
(d) it is the method for the elimination of excess of cholesterol from the body;
(e) it is concerned with the metabolism and excretion of bile pigments;
(f) it forms the excretory vehicle for certain drugs, toxins and transition metal ions.
[C:532]

636 Cholesterol gall stones are white or pale yellow stones with a high proportion of cholesterol and a low $Ca^{2+}$ content. The stones have formed because of some change in the colloid stability of cholesterol-lecithin-bile salt micelles in bile. Pigment gall stones are small hard green-black stones that are mainly bile pigment. They are found sometimes in patients with chronic haemolytic conditions. Although they contain a little $Ca^{2+}$, they are not usually radio-opaque. Mixed gall stones are the commonest type. They are hard, dark brown stones containing bile pigments, cholesterol, protein and sometimes sufficient $Ca^{2+}$ to give X-ray shadows.
[I:300]

# BLOOD AND URINE

## A  The blood clotting process

637 The 'coagulation vitamin' is vitamin K, which is a substituted 1,4-naphthaquinone. There are 3

main variants, all of which possess a methyl substituent at the 2-position of the quinone ring. Vitamin $K_1$ variants (phylloquinones) are of plant origin and have a long alkenyl side chain (containing one double bond) at the 3-position of the quinone ring. Vitamin $K_2$ variants (menaquinones), synthesized by bacteria, are similar to phylloquinones, but the alkenyl side chain possesses several double bonds. Vitamin $K_3$ (menadione) has no substituent at the 3-position. It and similar menadiones are synthetic substances. Whilst the natural vitamins $K_1$ and $K_2$ are fat-soluble vitamins, some of the simpler synthetic compounds have appreciable solubility in water.
[C:120]

638 Vitamin K deficiency is unusual, but can occur (a) in the newborn, (b) during treatment with neomycin or other drugs that suppress intestinal flora and (c) in conditions, such as obstructive jaundice, coeliac disease and ulcerative colitis, where fat absorption is impaired.
[C:120]

639 Dicoumarol and warfarin are coumarin derivatives with a structural similarity to menadione. They are anticoagulants possessing the ability to inhibit the activity of vitamin K (antivitamin K activity). Use has been made of the long term activity of these compounds in the prevention of further clotting following a cardiac embolism or in other thrombotic conditions.
[C:122]

640 Collected blood is added immediately to heparin, citrate, fluoride or to other suitable anticoagulants to prevent clotting before the plasma is separated from cells. Heparin is a large molecule containing long polysaccharide chains with numerous negative charges. It is normally present in blood at very low concentration, but its concentration increases during anaphylactic shock. One of its actions is the inhibition of thrombin formation by the activation of the protein, antithrombin III, present in plasma. The actions of both citrate and fluoride *in vitro* are probably related to the sequestration of $Ca^{2+}$ ions, which are necessary for the formation of active clotting factors. Citrate forms a calcium

complex and fluoride forms $CaF_2$ (of very low solubility).
[G:926]

641 Thrombocytes (platelets) adhere to exposed collagen fibres, to the edges of damaged blood vessels, to various foreign surfaces and to other platelets (if platelet stickiness has been increased by a suitable agent). This agglutination produces a platelet clot that not only acts as a focus for blood clot formation, but liberates various factors that are essential for the overall haemostatic process. Released serotonin has vasoconstrictor activity, platelet-derived phospholipid is required for the activation of protein clotting factors in the blood, and possibly other liberated substances assist fibrin formation and stabilization. When the platelets start to break down, they release a clot retraction factor that shrinks the clot in preparation for fibrinolysis. Thus, platelet response is complex and of vital importance to the whole clotting process.
[C:547]

642 Fibrinogen is a high molecular weight $(3.3 \times 10^5)$ protein of elongated shape (length/diameter = 20/1). It contains 6 polypeptide chains, 6 chain ends being located in the centre of the protein structure. Peptide segments at both ends of the elongated structure possess several negative charges at pH 7.4 and these charges help to prevent the coalescence of several fibrinogen structures. When thrombin is produced in plasma, it effects the hydrolytic cleavage of 2 pairs of peptides (fibrinopeptides A and B) from the ends of each fibrinogen structure. This removes the large negative charges and association of the structures develops rapidly to produce a 3-dimensional network of fibrin. The enzyme required for the fibrinogen → fibrin process is thrombin.
[C:547]

643 The fibrin network, when first produced, is soft and readily deformable. It is strengthened by the formation of covalent cross-links between lysine and glutamine side chains in neighbouring poly-

194

peptides as shown:

Fibrin                              Fibrin
chain                               chain

$$-CH_2-CH_2-CO-NH_2 + H_3\overset{+}{N}-CH_2-CH_2-CH_2-CH_2-$$

Fibrin
stabilizing
factor $\longrightarrow NH_4^+$
(FSF)

$$-CH_2-CH_2-CO-NH-CH_2-CH_2-CH_2-CH_2-$$

[C:548]

644 As healing of the blood vessel proceeds, the clot gradually shrinks to about half its original size under the action of a clot retraction factor. A complex sequence of activating processes gives rise to the formation of fibrinolysin from an inactive precursor. The proteolytic enzyme, fibrinolysin, then hydrolytically breaks down the fibrin matrix.
[G:927]

645 The precursor of thrombin is prothrombin, a protein normally present in plasma. The formation of thrombin requires the presence of $Ca^{2+}$ ions, platelet phospholipid, and 2 proteins. These proteins are proaccelerin (factor V) and the activated form of Stuart–Prower factor (factor Xa).
[C:548]

646 The events giving rise to the development of the fibrin clot are often divided into 3 processes:
(a) the formation of thromboplastic activity,
(b) the conversion of prothrombin into thrombin,
(c) the conversion of fibrinogen into fibrin.
Tissue damage, in the presence of $Ca^{2+}$ ions, triggers off a sequence of processes in which a set of protein factors, present at low concentration in plasma (or released by damaged tissue), become activated in turn. Each activated protein factor is the enzyme catalysing the activation of the next enzyme precursor. Operation of this enzyme cascade gives rise to the formation of

thromboplastic activity (the enzyme activity needed to convert prothrombin into thrombin).
[C:549]

647 Classical haemophilia is a sex-linked inherited disease that arises from failure to synthesize the coagulation factor VIII (antihaemophilic factor A) and consequently failure to develop thromboplastic activity.
[F:176]

648 Synthesis of prothrombin occurs in the liver and requires the presence of vitamin K. The vitamin is a cofactor required for the $\gamma$-carboxylation of glutamic acid residues present in the 'pro' portion of the molecule. The presence of $\gamma$-carboxyglutamic acid residues act as $Ca^{2+}$ binding sites, the association with $Ca^{2+}$ being a necessary stage in the conversion of prothrombin to thrombin. Vitamin K is also required for $\gamma$-carboxylations necessary in the synthesis of clotting factors VII, IX and X.
[F:175]

## B  Serum proteins

649 The main functions of various serum proteins may by summarized as:
  (a) controlling the distribution of water between intra- and extravascular spaces;
  (b) transporting vitamins, hormones, ions, etc.;
  (c) protecting (by the action of circulating components of the immune system);
  (d) acting as a source of nutrient protein for tissues;
  (e) possessing hormone and enzyme activity;
  (f) contributing (to a small extent) to the buffer action of blood.
  [I:305]

650 Electrophoresis on a cellulose acetate strip is the process used routinely for the separation of the main serum protein fractions. Various immunoassays, such as radioimmunoassay, electroimmunoassay and immunodiffusion, nephelometric methods and special activity measurements are available for the assay of many individual proteins.
[G:904]

651 (a) $\alpha_1$-Antitrypsin is a protein that inhibits proteolytic enzymes. Uninhibited protease activity can produce tissue breakdown and, in particular, causes liver and lung damage. Deficiency of $\alpha_1$-antitrypsin may produce cirrhosis in a child or emphysema in a young adult.

(b) Haptoglobin is a protein with the ability to bind to haemoglobin and so prevents the loss of haemoglobin (and its breakdown products) liberated in haemolytic conditions.

(c) Inactive complement proteins are synthesized by macrophages. These proteins are activated by antigen–antibody complexes and bacterial toxins. A reaction sequence is triggered and this gives products that attract phagocytes, cause local thrombosis, and increase vascular permeability.

[I:313, 314]

652 An immunoglobulin molecule contains 4 polypeptide chains, 2 long (or heavy) H and 2 short (or light) L, the chains being linked by disulphide bonds in the neighbourhood of a central 'hinge' region. The molecule has a shape that looks like the letter Y (or T), the 2 short arms each consisting of a light chain linked to part of a heavy chain. The ends of the short arms have a very variable composition and each comprises an antigen-combining site. Carbohydrate is associated with the H chains.
[I:315]

653 Immunoglobulins can be divided into 5 main groups, usually called IgG, IgA, IgM, IgD and IgE. These differ in their composition and type of antibody function. The IgM structure is much larger than the others and is made up of 5 subunits, each having the Y-type structure.
[I:318]

654 $\alpha_1$-Antitrypsin is present in the $\alpha_1$-globulin, haptoglobin and $\alpha_2$-macroglobin are in the $\alpha_2$-globulin, transferrin is in the $\beta$-globulin and IgG is in the $\gamma$-globulin fraction.
[I:308]

655 Hypoalbuminaemia might be caused by:
(a) liver disease reducing albumin synthesis;
(b) the nephrotic syndrome, protein-losing

enteropathy, and psoriasis or skin burns giving rise to large serum protein loss;
(c) increased protein catabolism that occurs in several diseases;
(d) protein malnutrition or malabsorption;
(e) increased synthesis of other serum proteins being offset by reduced albumin synthesis;
(f) overhydration resulting in haemodilution.
[I:309]

656 The classical abnormal electrophoretic patterns for the 4 conditions are:
(a) the nephrotic syndrome:
albumin (low), $\alpha_1$ (low), $\alpha_2$ (high), $\beta$ (normal), $\gamma$ (low);
(b) cirrhois:
albumin (low), $\alpha_1$ (low), $\alpha_2$ (normal), $\beta$ (normal, but not separating from $\gamma$), $\gamma$ (high);
(c) hypogammaglobulinaemia:
albumin (normal), $\alpha_1$ (normal), $\alpha_2$ (normal), $\beta$ (normal), $\gamma$ (low);
(d) monoclonal gammopathy:
albumin (normal), $\alpha_1$ (normal), $\alpha_2$ (normal), $\beta$ (normal), $\gamma$ (only a single short high peak).
[I:312]

## C  The kidney and urine

657 The glomerulus is a filtering membrane that separates blood from an ultrafiltrate containing all the small ions and molecules of plasma in their respective plasma concentrations, but containing only traces of protein. The glomerular filtrate (GF) is produced at a rate of about 120 ml/min and depends on a hydrostatic pressure gradient across the membrane.

Active processes in the proximal tubule cells effect the reabsorption from the GF of glucose, amino acids and other substrates required for metabolic processes. About 70% of the $Na^+$ is actively reabsorbed, accompanied by anions, and about 70% of the water. Hydrogen ions are secreted into the lumen of the proximal tubule.
[I:1]

658 The immediate causes of a reduced GFR are (a) reduction in the filtration pressure across the membrane, and (b) glomerulonephritis increas-

ing the membrane's resistance to the filtration process.
[I:12]

659 A process of countercurrent absorption of $Na^+$ and $Cl^-$ from the ascending limb of the loop of Henle to the descending limb leads to high osmolality at the bottom of the loop and to hypo-osmolal fluid entering the distal tubule. The osmotic gradient is required for vasopressin (ADH) to control further reabsorption of water in the distal tubule and collecting duct. The distal tubule also finally adjusts the $Na^+$, $H^+$ and $K^+$ concentrations by reabsorption of $Na^+$ and excretion of $H^+$ and $K^+$ ions.
[G:1066]

660 The onset of mainly tubular dysfunction can be caused by damaging effects of ingested toxic materials (e.g. large metal ions) or toxic damage to the tubules from substances present because of metabolic disorders such as hypercalcaemia, galactosaemia, hyperuricaemia and Wilson's disease. Both glomerular and tubular dysfunction tend to lead to generalized renal failure.
[I:14]

661 Many components of the GF are reabsorbed, but have a specific maximum possible rate of tubular reabsorption. If the flux of solute in the GF exceeds this $T_m$ value then the unabsorbed excess will be excreted in the urine. Substances such as urea, uric acid and creatinine have low $T_m$ values, whilst amino acids and glucose have high $T_m$ values.
[G:1065]

662 Amino acids are present in the GF at their plasma concentrations, but normally are reabsorbed almost entirely (in the proximal tubule) so that normal urine contains only traces of amino acids. Abnormally high urinary concentrations may arise for two reasons:
(a) an abnormally high plasma concentration of one or more amino acids may result in passage of these into the tubules at rates that exceed the particular $T_m$ values so that an overflow aminoaciduria results;
(b) if the rate of entry of amino acids into the GF

199

is normal, but a defect exists in the re-absorption mechanism, then one or more amino acids will appear in the urine, this being a renal aminoaciduria.

Specific aminoacidurias where one (or a few related) amino acids are excreted in the urine are invariably caused by an inherited defect.
[I:356]

663   In the Fanconi syndrome there is a non-specific aminoaciduria with excessive urinary excretion of other substances, such as carbohydrates and phosphates. This syndrome is usually caused by tubular cell damage.
[I:356]

664   Cystinuria is a congenital defect in tubular re-absorption of cystine and the basic amino acids, arginine, lysine and ornithine. This condition is often benign, but may lead to the formation of cystine stones.
[I:357]

665   Most renal stones (80–90%) consist largely of calcium phosphate and calcium oxalate. About 10% of the cases have stones that consist mainly of uric acid and a small proportion of the cases have cystine stones. Very rarely, the stones consist mainly of xanthine.
[I:24]

666   The following reaction system is present in both proximal and distal tubular cells:

$$CO_2 + H_2O \underset{\text{anhydrase}}{\overset{\text{carbonic}}{\rightleftharpoons}} H_2CO_3 \longrightarrow H^+ + HCO_3^-$$

(from blood)

to lumen    to blood

The GF entering the proximal tubule contains $Na^+$ and $HCO_3^-$ at the plasma concentration, but the $H^+$ ions that pass into the lumen from the reaction scheme convert $HCO_3^-$ ions into $H_2CO_3$, which can then supply $CO_2$ to the cells to supplement the $CO_2$ obtained from the plasma. Transport of $Na^+$ takes place from the lumen, through the cells to the circulation so that electrical neutrality is maintained. The overall effect is to retain $Na^+$ and $HCO_3^-$ ions.
[I:85]

667 The same reaction sequence, given in the previous answer, exists in the distal tubular cells, but the $H^+$ ions entering the lumen are largely removed by 2 processes:
(a) $H^+ + HPO_4^{2-} \rightleftharpoons H_2PO_4^-$
(b) $H^+ + NH_3 \rightleftharpoons NH_4^+$
The $HPO_4^{2-}$ ions are present in the urine and the conversion of a proportion of these to $H_2PO_4^-$ ions increases the titratable acidity of urine. Ammonia for process (b) is produced specially in the tubular cells by the action of glutaminase on glutamine. The enzyme is activated by the acidic pH.
[I:88]

668 The main solute components of normal urine are urea, creatinine, uric acid, and ammonium, sodium, calcium, chloride, phosphate and sulphate ions. There are very many other compounds present at low concentration.
[G:1076]

669 Reagent strips and also tablets (e.g. by Ames) are available for the routine testing of fresh urine. Test strips are used, for example, to test for protein, glucose, ketone bodies, bile pigments, phenylpyruvic acid and blood. The Ames clinitest is a tablet-based test method for the semi-quantitative assay of reducing substances and the Ames acetest is a tablet method for combined acetoacetate and acetone. Other tablet tests are available. Combined strip tests (e.g. Ames Labstix) give results for more than one urinary component.
[I:206, 208, 331]

# BIOCHEMISTRY OF TISSUES

## A Connective tissue (general)

670 In general, connective tissue contains:
    (a) specific fibroblasts, such as osteoblasts, chondroblasts and odontoblasts;
    (b) other cells (frequently with a protective action), such as leucocytes, macrophages, plasma cells, mast cells and fat cells;
    (c) collagen and/or sometimes other fibrous protein, e.g. reticulin and elastin;
    (d) ground substance, which is the encompassing matrix and contains mucopolysaccharides (usually associated with protein as proteo-

glycans), glycoproteins, phospholipids and water.

Mineralized connective tissue (e.g. calcified cartilage, bone, dentine) also contains forms of calcium phosphate and other minerals.

[G:1134, 1146]

671 Each chain contains about 1000 amino acid residues, with glycine as every third residue along the chain. Proline (12.5%), alanine (11.5%) and 4-hydroxyproline (9%) together make up another 33%, whilst the 2 acidic amino acids add 12.5% and 3 basic amino acids add a further 8% to the total number of amino acids in a chain. The 3 basic amino acids include δ-hydroxylysine. Hydroxyproline and hydroxylysine are not found in proteins outside the collagen–reticulin–elastin group. Tryptophan and cysteine are virtually absent.

[F:186]

672 The tropocollagen molecule contains 3 single collagen polypeptide chains, each chain being present as a left-handed helix with about 3 residues per turn. The 3 chains fit together to form a tightly wound triple helix held together by hydrogen bonds. The tightly coiled internal structure of a tropocollagen molecule is only possible because of the high content of glycine (enabling the chains to fit closely together) and of proline and hydroxyproline (imposing the required sharp changes of direction and also local rigidity in the 3 individual chains).

[F:188]

673 Each tropocollagen molecule is like a long thin rod with little flexibility. These molecules are arranged in parallel linear bundles often of cylindrical form, up to about 100 nm in diameter and of indefinite length. The individual tropocollagen molecules are staggered and the assumption is often made that each molecule overlaps its neighbour by about one-quarter of its length, leaving a gap of about 40 nm between the tail of one molecule and the head of the next in line. This allows reasonable flexibility to the fibril and allows it to adopt a curved path. The hydrophilic character of the structure preserves the extended form in an aqueous medium. Parallel molecules of tropocollagen are held together, not only by hydrogen bonds, electrostatic forces

and, to a lesser extent, hydrophobic association, but also by some strong covalent bridges. These bridges form from the oxidation of some lysine side chains to give aldehyde groups

$$-(CH_2)_3.CH_2NH_2 \rightarrow -(CH_2)_3.CHO,$$

followed by condensation of the latter with un-modified lysine or hydroxylysine residues in neighbouring chains. Reduction of the $-CH=N-$ groups in the Schiff's bases produced by this condensation, gives rise to bridges of the type

$$-(CH_2)_3-CH_2-NH-CH_2-(CH_2)_3-.$$

[F:194]

674   Nascent polypeptide chains (protocollagen) are assembled on the ribosomal complexes of fibro-blasts and selected proline and lysine residues are hydroxylated, possibly before synthesis of each chain is complete. Protocollagen hydroxyl-ase requires the presence of $\alpha$-oxyglutarate, $Fe^{2+}$, ascorbic acid and $O_2$. The polypeptide chains, each about 20% longer than the individual tropocollagen chains, undergo glycosylation at some hydroxylysine residues before being ex-truded from the cells. Each tropocollagen triple helix is then formed from the appropriate 3 single chains, this process involving the formation of a triple-helical procollagen molecule, followed by the removal of 6 excess chain lengths. Fibrils then form and covalent cross-links are forged.
[F:193, 196]

675   The triple helix of tropocollagen is a stable struc-ture, but collagenases are present in tissues and turnover of collagen occurs, its rate depending on the tissue, including factors such as its age and whether injury is present. Collagenase activity requires close control. Collagenases are inhibited by the protease inhibitors $\alpha_1$-antitrypsin and $\alpha_2$-macroglobulin.
[F:198]

676   Proteoglycans are carbohydrate–protein com-plexes in which the carbohydrate content is much greater than the protein content. A single specific glycosaminoglycan (mucopolysaccharide) chain may possess hundreds, or even thousands, of monosaccharide residues with at least half that

number of negative charges. Many (perhaps 50–100 or more) of these chains are linked to serine residues on a central polypeptide chain. The carbohydrate chains of these giant molecules repel each other and, in water, the molecule occupies a large volume with the polypeptide acting as the central axis and the charged carbohydrate chains radiating like the bristles of a sweep's brush.

Glycoproteins are mainly protein with one or more short carbohydrate chains (2–20 residues) covalently attached to serine residues. The chains usually contain N-acetylhexosamine residues (also in proteoglycans), but they lack the uronic acids present in proteoglycans. The carbohydrate chains of glycoproteins often carry a specific array of negative charges.
[C:430, 438]

677 Glycosaminoglycan chains are made up of repeating disaccharide residues, each disaccharide residue containing an N-acetylhexosamine usually linked to a uronic acid (such as D-glucuronic acid or L-iduronic acid). In many mucopolysaccharides, the N-acetylhexosamine residues, and sometimes the uronic acid residues, have an –OH group converted into a sulphate ester so that, at physiological pH, both $-OSO_3^-$ and $-COO^-$ substituents may be present in great numbers along the chain.

Useful reference structures are

N-acetyl-β-D-glucosamine          β-D-glucuronic acid

In hyaluronic acid, N-acetyl-D-glucosamine is linked to D-glucuronic acid by a β $(1 \rightarrow 4)$ linkage and the latter residue is then linked to the next N-acetyl-D-glucosamine by a β $(1 \rightarrow 3)$ linkage.

Chondroitin 4-sulphate differs from hyaluronic acid in that N-acetyl-D-galactosamine 4-

sulphate replaces N-acetyl-D-glucosamine in each disaccharide unit.
[F:200]

678    Proteoglycan molecules, whilst helping to maintain the structural integrity of soft connective tissue, can hold large quantities of water and form a milieu suitable for the development of fibrous structures and for the free diffusion of nutrients to, and waste products from, the connective tissue cells. Collagen acts with proteoglycans and the other components of connective tissue to form an integrated structure with the properties necessary for the specific physiological role of the tissue.
[G:1152]

679    In contrast to collagen, turnover rates of mucopolysaccharide chains in the body are high. Hyaluronic acid and chondroitin sulphates appear to have half-lives in the range of 2–4 days in animals.
[G:1155]

680    Mucopolysaccharidoses are a group of inherited diseases with specific lysosomal enzyme deficiencies. In consequence, there is accumulation of mucopolysaccharide in tissues and urinary excretion of sulphated mucopolysaccharide chains. These diseases, in general, produce skeletal deterioration, corneal clouding, hepatosplenomegaly and mental retardation. The prognosis is usually poor. Hurler's syndome (gargoylism) is an autosomal recessive condition with deficiency of $\alpha$-L-iduronidase. The disease develops in late infancy and shows all the above clinical features. Morquio syndrome is an example of a mucopolysaccharidosis in which mental development is normal, but skeletal deformities are severe.
[A:174]

## B    Mineralization and the composition of bone and teeth

681    Bone is lined with stem cells and these can differentiate into mobile multinucleate osteoclasts capable of breaking down the organic matrix of bone and dissolving the bone material. Both stem cells and osteoclasts can develop into osteoblasts. The latter cells can synthesize the components of

the organic matrix and secrete vesicles containing mineral and alkaline phosphatase. An osteocyte is a mature osteoblast surrounded by newly formed bone which it is capable of remodelling. Osteocytes are connected by long processes that also link to the stem cells.
[A:986, 988]

682 Bone mineral contains very small crystals (about $15 \times 5 \times 5$ nm) of biological apatite, together with even smaller spheres of amorphous calcium phosphate. Biological apatite is essentially calcium hydroxyapatite $Ca_{10}(PO_4)_6(OH)_2$, but with some variation in the composition of the crystals. These crystals form at sites on collagen fibres and the total mineral content of the bone can rise to about 70%, with collagen contributing a further 17–20%. Ground substance and water make up the remainder.
[A:986]

683 Biological apatite minerals form small crystals with defects in the crystal lattice structure at the ionic level. The molar $Ca^{2+}/PO_4^{3-}$ ratio in $Ca_{10}(PO_4)_6(OH)_2$ is 1.67, but in biological apatite this ratio can be between 1.3 and 2.0. The X-ray diffraction patterns of bone mineral are poorly defined, but are sufficiently similar to those of the larger, better formed, and harder crystals of calcium fluorapatite to indicate that the crystal structures of these minerals are broadly similar. Biological apatite crystals contain carbonate (*ca.* 5%) and smaller proportions of other ions including magnesium, sodium, potassium, citrate, chloride and fluoride. These ions may occupy gaps or defects within the hydroxyapatite crystal lattice or be adsorbed onto the very large surface area of this submicrocrystalline material.
[H:188, 218]

684 Calcified cartilage, bone, dentine and cementum are mesodermal in origin and are broadly similar, in that they contain the very small biological apatite crystallites and amorphous calcium phosphate in an aqueous organic matrix of which the main component is collagen. The cartilage contains a much higher proportion of hydrated proteoglycans (mainly derivatives of chondroitin sulphate A and dermatan sulphate) than the

other calcified tissues. Collagens from dentine and cementum are very rich in cross-linkages.
[H:204, 215]

685 Enamel is a unique calcium hydroxyapatite system that combines great hardness with a suitable degree of elasticity. Impact forces on enamel get transmitted to the softer dentine that lies below. Mature enamel contains about 95% of apatite crystals, about 4% of water and about 0.5% of organic components. The individual apatite crystals are about 10 times larger in *each* dimension than those in other calcified tissues and are packed together with only minute channels for water and other components. This system of numerous crystals forms the prisms that are recognized histological structures. The protein present in immature, and in mature, enamel does not consist of collagen.
[H:208]

686 Fluoride ions at low concentrations are adsorbed by, and then enter, the enamel structure converting some hydroxyapatite crystal lattice (particularly defective lattice) to the harder fluorapatite structure, which is less soluble in acid produced by bacterial action. The presence of fluoride ions also appears to assist the process

$$Ca_8H_2(PO_4)_6.5H_2O + 2Ca^{2+} \rightarrow$$
$$Ca_{10}(PO_4)_6(OH)_2 + 4H^+ + 3H_2O$$

that occurs in the growth of a hydroxyapatite crystal.
[H:263]

687 The solubility product $K_s$ of calcium hydrogen phosphate is defined by the equation

$$K_s = [Ca^{2+}] \times [HPO_4^{2-}]$$

It has been found that the same equation will also approximately govern the solubility of defatted powdered bone mineral in serum if the $K_s$ value is reduced to about one-quarter of its value for calcium hydrogen phosphate.
[H:185, 197]

688 The ionic product for free $Ca^{2+}$ and $HPO_4^{2-}$ ions in plasma exceeds the reduced $K_s$ value, referred to in the previous answer, by a factor between 1.5

and 3.0. This indicates that plasma is supersaturated with respect to calcium hydroxyapatite. However, crystallization can commence only if a suitable pattern of electrical charges is present on some structure and individual ions can assemble on this pattern to start the crystal lattice. Once started, the crystal structure can grow rapidly. The collagen fibril structure probably provides this arrangement of charges.

Two other factors are concerned with the localization of the mineralization process:

(a) the vesicles produced by osteoblasts probably contain $Ca^{2+}$ and $HPO_4^{2-}$ ions at concentrations greater than in plasma;

(b) pyrophosphate $(P_2O_7^{4-})$ present in plasma and in other extracellular fluids inhibits formation of the hydroxyapatite crystal lattice. In hard tissues, alkaline phosphatase hydrolyses the $P_2O_7^{4-}$ ion to give 2 non-inhibiting $HPO_4^{2-}$ ions.

[A:987]

## C  Muscle and nerve

689   A mature skeletal muscle cell can be several mm in length. It is enveloped by an electrically excitable membrane (the sarcolemma) and contains nuclei situated immediately beneath the sarcolemma. A sarcoplasmic reticulum exists in the cytosol and many mitochondria are usually present. Cytosol contents include glycogen, glycolytic enzymes, phosphocreatine and ATP. Striated muscle cells contain many parallel myofibrils composed of linearly arranged structural units called sarcomeres, each about 2.3 $\mu$m in length.
[G:1085]

690   A sarcomere contains parallel arrays of thin and thick protein filaments, the latter being twice the diameter of the former. Thick filaments contain myosin and thin filaments contain actin, tropomyosin and troponin. Cross-sections cut through the most dense regions show regular hexagonal arrays, with each thick filament having 6 thin filaments as its nearest neighbours and each thin filament having 3 thick filaments as its nearest neighbours. The degree of contraction of the muscle depends on the extent of longitudinal overlap of the parallel thick and thin filaments.
[F:815]

691 Myosin is a very large protein (molecular mass, $5 \times 10^5$) consisting of 2 identical chains (each of molecular mass $2 \times 10^5$) wound together as a double helix to form a long rod-shaped structure. At one end of this rod, the chains separate and each separated portion forms a globular region that possesses a site with ATPase activity and a site that binds actin. Two light polypeptide chains, each of molecular mass $2 \times 10^4$, are attached to each globular region and these chains may exert some fine control on the ATPase activity.

Actin appears to consist of 2 thick strands coiled as a double helix into a long structure and each strand is composed of a linear polymeric sequence of G-actin monomers. These monomers consist of globular proteins (molecular mass, $4 \times 10^4$).

Tropomyosin is a double-stranded helix that is wound around an actin fibril and can block myosin-binding sites on the actin. Troponin is a complex of 3 polypeptide chains labelled TnT, TnI and TnC. TnT binds to tropomyosin and TnI binds to actin. TnC molecules possess binding sites for $Ca^{2+}$ ions.
[F:819]

692 Muscle contraction occurs because of the sliding of thin filaments over thick filaments. The thick filaments consist of bundles of myosin rods with their globular regions (heads) pointing radially out of the filament. The centre region of the filament is 'bare' of these globular regions whilst the direction in which the molecules lie (polarity) is different on opposite sides of this region. The polarity of actin molecules in the thin filaments is, of course, complementary to that of the myosin molecules. Thus the polarity in both thick and thin filaments reverses in the centre of a sarcomere. When muscle is stimulated, myosin rods bend and the heads move outwards from the thick filament and attach to binding sites on the actin structure. The heads tilt, moving the thick filament relative to the thin filament and they are then released from the actin. Before stimulation occurs, ADP and $P_i$ are strongly bound to the myosin head, but are released on attachment of the head to actin. This release produces the change in conformation. Separation of the tilted head from the actin structure is brought about by

reaction with ATP. Hydrolysis of the attached ATP to $ADP + P_i$ completes the sequence. Each myosin molecule has 2 heads and many molecules are acting in concert along the filaments to effect the muscle contraction.
[F:821, 824]

693   Arrival of a nerve impulse at a motor end plate depolarizes the muscle fibre membrane and effects the rapid release of $Ca^{2+}$ ions from the sarcoplasmic reticulum within the cell. These $Ca^{2+}$ ions stimulate muscle contraction. They bind to the TnC polypeptide, resulting in a conformational change in tropomyosin that exposes myosin head binding sites present on the actin structure. On return to the resting state, an ATP-utilizing mechanism concentrates $Ca^{2+}$ ions in the sarcoplasmic reticulum.
[F:826]

694   Glucose, ketone bodies and fatty acids can act as the energy source and together with the breakdown of glycogen already stored in the striated muscle cells provide, in their metabolism, the source of ATP required for muscle action. However, the actual ATP content of a muscle cell would not in itself supply energy for even a second of muscular activity. The next store of high energy phosphate is phosphocreatine, creatine kinase catalysing the reaction:

$$ADP + Phosphocreatine \rightleftharpoons ATP + Creatine.$$

Protracted muscular work leads to the build-up of lactic acid, because of the temporarily inadequate supply of oxygen to the muscular tissue. Some oxygen is stored in muscle cells as oxygenated myoglobin.
[G:1097]

695   Acetylcholine is the diffusible molecule that acts as the impulse transmitter at junctions between nerve and striated muscle (motor end plates). Two molecules of acetylcholine then bind to a receptor site on the motor end plate, which in the resting state has a potential of about $-75$ mV. This binding produces a conformational change in the membrane that opens a channel and allows a rapid passage inwards of $Na^+$ ions (and a small opposite flow of $K^+$), this ion flux de-

polarizing the membrane and triggering an action potential in the muscle membrane.
[F:887]

696   Acetylcholinesterase catalyses the hydrolysis of acetylcholine, the reaction being:

$$(CH_3)_3\overset{+}{N}.CH_2.CH_2O.COCH_3 + H_2O \rightarrow$$
$$(CH_3)_3\overset{+}{N}.CH_2.CH_2OH + CH_3.COO^- + H^+$$

Acetylcholesterase, being an enzyme of great catalytic power, rapidly hydrolyses acetylcholine and quickly restores the membrane polarization.
[F:890]

697   Noradenaline (R = OH) and dopamine (R = H)

(displayed as their cations in the formula) are examples of catecholamine neurotransmitters. They are inactivated by the action of either catechol-O-methyltransferase, which transfers a methyl group from S-adenosylmethionine to the 3-OH group in the benzene ring, or a monoamine oxidase, which oxidizes the $-\overset{+}{N}H_3$ to a $-CHO$ group.
[F:895]

698   $\gamma$-Aminobutyrate (GABA) acts to increase the permeability of postsynaptic membranes to $K^+$ ions and thus inhibits the triggering of action potentials, i.e. it is an inhibitory transmitter. Transamination inactivates GABA by converting it into an aldehyde, which then undergoes dehydrogenation to give succinate:

$$GABA \rightarrow OCH-CH_2-CH_2-COO^- \rightarrow$$
$$^-OOC-CH_2-CH_2-COO^-$$

[G:1118]


# METABOLISM OF FOREIGN COMPOUNDS

## A   General principles, absorption and excretion

699   In general, foreign substances with high solubilities in membrane lipids are absorbed, enter

211

cells and cross membranes more readily than those with low lipid/water partition coefficients. [D:444]

700 The breath is a major route for the excretion of diethyl ether and other highly volatile materials.

701 Many lipid-soluble materials are readily absorbed by these routes. Thus:
   (a) anaesthetic gases and vapours of industrial solvents are absorbed by the lungs;
   (b) low polarity solvents and oils are absorbed through the skin, so that some drugs in an oil base may be appropriately administered by rubbing into the skin.
   [B:207]

702 The statement contained in the question is only correct for weak acids $HB^{z+}$ for which $z = 0$. Such weakly acidic drugs would be present in the un-ionized form at the low pH of stomach contents but mainly in the ionized form in alkaline intestinal fluids. Then, because of the higher lipid solubility of the ionized form, absorption will occur from the stomach.
   [A:1246]

703 Assuming that the base itself is uncharged and is not too weak, then it will be present in the positively charged (protonated) form in the stomach. In this form it will not be readily absorbed from the stomach. In the alkaline intestinal fluid it will be in the un-ionized form and absorbed much more readily.
   [A:1246]

704 Neomycin is sufficiently strongly basic to be mainly in the charged (protonated) form, not only in gastric fluid but in intestinal fluid as well. Hence, it is poorly absorbed and high intestinal concentrations can be maintained.
   [C:539]

705 Excretion of an acid or base will occur much more readily when in the ionized form and thus urinary pH may influence the excretion of weak acids and bases.
   [G:1078]

706 Structural modification of a foreign substance in

the liver often detoxifies, but in other cases it can give rise to a more toxic product.
[A:1250]

707 Examples of (a) reducing and (b) increasing activity are respectively:
   (a) phenobarbitone is metabolically inactivated by hydroxylation of the benzene ring to give *p*-hydroxyphenobarbitone;
   (b) molecules of the azo dye prontosil are split (by reduction) to give, as one product, the active antibacterial agent sulphanilamide.
[A:1250]

## B  Phase 1 reactions

708 Phase 1 or 'non-synthetic' reactions are modifications such as hydrolysis, reduction, dealkylation, and oxidation. These reactions often precede reaction with conjugating molecules such as carbohydrates and amino acids. Phase 1 reactions often increase the hydrophilic character of the molecules. A phase 2 or 'synthetic' reaction often follows a phase 1 reaction and may involve the formation of a water-soluble and often negatively charged conjugate.
[A:1250]

709 Enzymes involved in many phase 1 reductive and oxidative processes are microsomal. Hydrolytic enzymes are found in the soluble fraction of liver cells.
[A:1251, 1253, 1250]

710 Microsomal aromatic hydroxylations (a), and deaminations and *N*-dealkylations (b), involve mixed function oxidase reactions of the types:
   (a) $R-H + O_2 + NADPH + H^+ \rightarrow R-OH + NADP^+ + H_2O$

   (b) $R-CH_2-NHR' + O_2 + NADPH + H^+$
   $$\downarrow$$
   $R-CHO + H_2NR' + NADP^+ + H_2O$

The processes are complex, transference of the oxygen requiring the presence of cytochrome P450.
[A:1255, 1258, 1256]

## C Phase 2 reactions

711 The addition of a very hydrophilic and often charged ligand to a xenobiotic (or xenobiotic modified by a phase 1 reaction) produces a product readily excreted in the urine (and possibly in the bile).
[A:1260, 1263]

712 Benzoic acid forms a more strongly acidic conjugate (hippuric acid) with glycine. Phenylacetic acid forms a more strongly acidic conjugate with glutamic acid. In the case of benzoic acid the process is as follows:

[G:1080, 740]

713 Any alcohol, phenol, primary amine and carboxylic acid may react with UDP-glucuronic acid to give a derivative of D-glucuronic acid with a $\beta$-linkage at the 1 position. These derivatives will be ethers, ethers, secondary amines and esters respectively.
[A:306, 1260]

714 Sterols and some other alcohols, phenols and primary amines can undergo conjugation to form sulphates.
[A:1263]

715 Taurine derivatives, ribosides and glucosides are examples of other conjugates. Some halogenated hydrocarbons can react with the –SH group of glutathione to give, after hydrolysis of peptide bonds and acetylation, conjugates of mercapturic acid. For example, bromobenzene gives the following mercapturic acid derivative:

$$Br-\langle \bigcirc \rangle -S-CH_2.CH.NH.COCH_3.$$

with a COOH group on the CH.

[G:61, 710]